The Magnificent Foragers

of Natural History
Smithsonian Institution

Smithsonian Exposition Books
Washington, D.C.
1978

Distributed to the trade by
W. W. Norton & Company
New York, N.Y.

Contents

The Smithsonian Institution
Secretary, S. Dillon Ripley

The National Museum of Natural History
Director, Porter M. Kier

Consulting Editor for
The Magnificent Foragers
Edward S. Ayensu

Writer, Thomas Harney (except where noted)

Smithsonian Exposition Books
Director, James K. Page, Jr.
Senior Editor, Russell Bourne

Staff, The Magnificent Foragers:
Alfred Meyer, Editor
Caren Keshishian, Picture Editor, Research
Amy Donovan, Copy Editor
Ann Beasley, Production Editor
Patricia Upchurch, Assistant

Thomas Hoffman, Business Manager

Bettie Loux Donley, Managing Editor

Design: Michael David Brown, Inc.
Michael David Brown, Art Direction
Assisted by Beth Haney, Jeanne C. Kelly, Helen Vickers, Turk Winterrowd

William H. Kelty, Marketing Consultant

Separations and Engravings:
    Lanman Lithoplate, Inc.
Typography: Carver
    Photocomposition, Inc.
Printing: Rand McNally and Company

Library of Congress Number 78-61066
ISBN 0-89599-001-6

Jacket photograph: Paul Spangler and field workers collecting beetles in Ecuador.

Foreword

The father of the National Collections was Spencer Fullerton Baird. He was a young professor of natural history at Dickinson College, Carlisle, Pennsylvania, in 1846 when the Smithsonian Institution was founded as "an establishment for the increase & diffusion of knowledge among men." Under the terms of the charter it was to be responsible for objects of natural history acquired by the United States. Baird wrote to Joseph Henry, the Institution's first Secretary, and won the job of managing what was to become a national storehouse.

When he arrived in Washington, D.C. in 1850 to assume his new duties, he brought with him the enormous collection he had accumulated since childhood: two freight cars full of bird and mammal skins, bones, fossils, bottled fishes, and reptiles. It was the Institution's first large natural history collection.

Even more valuable than this collection, however, were Baird's plans for putting such materials to use. In the mid-19th century, natural history specimens were still collected chiefly for exhibition as exotic curiosities. Baird believed, however, that for the sake of knowledge it was important to describe and catalog them, whether they were fossil fishes or Indian pots. He wanted the Institution to become a place where scientists and serious students could come to inspect and study representative samples of the flora, fauna, geology, and anthropology of America and the world.

Baird arrived on the job just at a time when America west of the Appalachian Mountains was being settled and prospected for the first time. The Government sent out survey and exploring parties whose duties included recording data about the West's vegetation, animals, geological formations, and Indians. The Institution became the scientific adviser to those expeditions, and it fell to Baird to select the men who were going to do the scientific work. He issued them the necessary equipment, trained them in its use, saw to it that they knew what was wanted and how to get it, and inspired them with his enthusiasm.

The Smithsonian's reward was custody of the crates full of specimens brought back to Washington. Within 10 years the Institution possessed the largest natural history collections in America. A vast amount of what was brought back was new to science. Baird, who became the leading authority of his time on birds, fishes, and reptiles, labored mightily to catalog and publish descriptions of this material. To help him classify and manage it, he brought young scientists to the Smithsonian, many of whom subsequently became world renowned.

Today's scientific staff descends from this original group. It is equally gifted and dedicated to the collections, which by now fill every available niche of space in the National Museum of Natural History. One reason for this plethora is that field explorations have continued since Spencer Baird's time. Every continent and large island group in the world has been visited by Smithsonian scientists, who in a real sense participate in an intellectual migration back and forth between the Museum and their field sites. They go to the field in part for specific examples—whether tangible like fossils, or abstract like notes of a conversation with a tribal chief—that will lead to more general conclusions. The field, or nature, keeps the scientists honest, as it were. They must always return to it, for it contains the raw materials of their research. By the same token, however, they must always return to the Museum which, in a certain sense, keeps nature honest. For, with the help of the collections, scientists can here perceive broader patterns of change in the flux of nature than are ordinarily visible in the field.

It is stimulating, productive rhythm, this migration. And it is undertaken by "magnificent foragers," indeed. They forage for knowledge, and in the pages that follow, you will become privy to both the migration and the "forage" that is its goal.

Porter M. Kier
Director
National Museum of Natural History

I.   The Tradition

Behind the working scientists at today's National Museum of Natural History stands a multitude of greats—the "Grand Old Men and Women of Smithsonian Science."

The Tradition

By John Sherwood

I found it in a legendary land
all rocks and lavender and tufted grass,
where it was settled on some sodden sand
hard by the torrents of a mountain pass.

I found it and I named it, being versed
in taxonomic Latin; thus became
godfather to an insect and its first
describer—and I want no other fame.

—"On Discovering a Butterfly,"
by Vladimir Nabokov

It all begins, for many natural scientists, in childhood when they are stricken on a magic day for a magic reason; when they curiously take the time for that second, longer look at a butterfly in the back yard. The third time around they may use a pair of field glasses, followed by the chase, the capture, and the mounting of the specimen for its delicate, showy beauty.

Next, perhaps, comes the purchase of W. J. Holland's *The Butterfly Book* to help them identify what it is they have caught. A fine-mesh insect-collecting net is added to aid in the determined, headlong pursuit through brambles and bush. The naturalist's natural collecting and hoarding instinct soon takes over, and the boy who would be president of the family firm has lost his heart forever to the wilderness.

The naturalist's almost-cursed, almost-bewitched drive to collect is a never-ending quest for knowledge. Triggered by the theory of evolution, they are urged on by an insatiable, compelling impulse to learn all there is to know in a specific field that is eventually narrowed down simply because lifetimes are, indeed, too short to cope with the subject at hand. They evolve into experts who learn more and more about less and less.

Department of Anthropology photo, taken in the early 1900s.

The Magnificent Foragers

Alexander Wetmore

For people like Alexander Wetmore, a past Secretary of the Smithsonian Institution and the Museum's most famous ornithologist, it all began when his mother gave him Chapman's *Handbook of Birds of Eastern North America.* In 1978, at the age of 91, the infirm Wetmore still has in his vast library his first nature diary and field notebook he started at the age of 8. The first page, dated November 1894 from Palmetto, Florida, goes like this: "There are a great many pelicans around. The pelican is a great big bird that eats fish. It has a pouch in which it keeps the fish until he wants to eat them."

By 1899, bird-watching with "Papa" and his dog Wags, he spotted a male goldfinch, and "here is the song: *Sweetsy sweeter sweet toosy toosy pe-e-e-e seoo per-chic o-ree swee swee per chic-o-ree.*" His first published work appears in the October 1900 issue of "Bird-Lore." It is "By Alick Wetmore (age, 13 years), North Freedom, Wis." and entitled "My Experience with a Red-Headed Woodpecker."

Since then, Wetmore has added 80 new birds to science; published his magnum opus, three massive volumes on *Birds of the Republic*

Wetmore at age 15

The Tradition

of Panama, and contributed more to the field of avian paleontology than any other person.

"One cannot help but be humbled to think that this is but a fraction of his total scientific output," says a colleague, Storrs L. Olson. His stupendous work in ornithology is staggering, and along the way to describing 189 species and subspecies of birds previously unknown. Wetmore—to get an idea of the respect other scientists had for him—had named after him: a sponge, a bat, a cactus, a deer, a glacier, an extinct eagle, 16 modern species and subspecies of birds, five mollusks, seven reptiles and amphibians, two fishes, nine insects, a canopy bridge in the Bayono River Forest in Panama, and a stork—which is particularly interesting because the tall (6 feet, 3 inches), handsome and graceful Wetmore himself resembles a stork.

A prime example of his absolute devotion to natural science was when a botfly larva burrowed into his leg during a bird-collecting expedition to the jungles of Panama. Wetmore, knowing that a colleague back at the Smithsonian was very much interested in the life cycle of this tropical fly, decided not to remove it. Back home, he allowed it to fester while the excited fly man kept watch on the development of the larva.

It must have been a very painful experience. The larva lodges on a nerve, and every time it turned it brushed upon the nerve. When the moment came for Wetmore to "give birth," he calmly excused himself from a ceremony and dashed off to his friend, the fly man. The winged specimen was successfully delivered and is probably somewhere in the collection today. That, my friends, is devotion.

* * *

To this day the serious youngster with the arrowhead collection, the botanical collection, the mineral collection, the bird's nest collection, is encouraged by the curators who themselves started out that way. The older fellows have never been able to turn their backs on their own childhoods, and in this way the torch is passed.

Jason Weintraub, a fledgling Detroit student lepidopterist interested in specializing in genetics, spent part of the summer of 1977 working at the Smithsonian. Jason, 17, came there because of Donald R. Davis, 43, head of the Department of Entomology, and Dr. Davis came there because of J. F. Gates Clarke, 73, now a research associate and retired former head of the department.'

Davis was impressed with Jason's work much as Dr. Clarke had been impressed with young Davis's work. "Jason wrote us, saying that he would like to work at the Museum for the summer," Davis recalls. "I wrote back, telling him what I thought of his work with lepidoptera, but I also had to tell him that we had no funds for this sort of thing." Jason wrote back, volunteering to work for nothing—the same arrangement that Clarke has today.

So Jason—son of Dr. Gerald Weintraub, an internist, and Dr. Rosalyn Weintraub, a pediatrician and dermatologist—found himself in the Museum of Natural History cataloging leaf-mining moths and learning to dissect specimens under the microscope and to mount them on slides. For the summer of 1978 Jason has an invitation to work at the Smithsonian Tropical Research Institute at Barro Col-

J. F. Gates Clarke

Mary Jane Rathbun

orado Island in the Panama Canal Zone. He hopes to study the population genetics of a group of Central American butterflies.

Jason, like Davis, became interested in natural science as a very young boy by way of dinosaurs. Davis used to dig up buried cows during his boyhood in Oklahoma, pretending that they were dinosaur bones. He would even try to put them together—you know, like they do at the Smithsonian. Then he went on to snakes, but household panic set in when the snakes occasionally escaped. Insects came next. They were revolting things, of course, but not as bad as snakes. And besides, they were dead.

Clarke, a trained pharmacist who left that field to study and collect moths and butterflies, saw the 17-year-old Oklahoma kid's lepidoptera collection at a science fair and nearly fell over. "It was an astounding collection," Clarke remembers. "I asked him who made his identifications. He said that he did. I didn't believe him at first. He became, of course, one of our star pupils."

* * *

Haunted by time, the mature naturalists press on, squirreled away in incredibly cluttered offices that hide them from view, like insects under rocks. They just don't have the *time* to do it all, and they prefer to be alone, forgetting the clock and hunger and comfort, often forgetting themselves and their families as well. They are driven men and women who will even work without pay.

Hung up and strung out on subjects so exotic that, in some cases, only a half-dozen others in the world can understand exactly what they are saying in publications, they chase the puzzling histories of their illusive and uncommunicative subjects. Be it lice or toads or crabs or mollusks or whales or brachiopods or human skeletons, the Natural History Museum has an expert in the field. They thrive, they live for their life's work, and compulsory retirement at the age of 70 means absolutely nothing to the most devout of them. More than anything, it means complete freedom to devote even more hours to their favored task as time struggles to take them away from it.

Waldo LaSalle Schmitt ("Uncle Waldo") was a case in point. Known as "The Shrimp Man," the call of crustacea lured him back to his office almost up until the time he died in 1977 at the age of 90, two decades after his "retirement." A world-famous marine zoologist, he was typical of the old order of Smithsonian naturalists, the all-around biologist who was interested in everything and who collected for his colleagues as well as for himself.

Dr. Schmitt, toward the end of his years, wore vintage World War I bow ties and round, gold-rimmed spectacles, yet he understood the scientific age in which he lived. He was a last survivor of the old days when Museum men always looked old, even if they were young.

They invariably dressed in black then, and were as dusty and as creaky as the antique exhibit cases. After World War II, however, the G.I. Bill changed all that. Scientists were getting their "papers" (Ph.D.'s). They had families. The pay was no longer a pittance. They could afford things like sport coats and homes and automobiles.

The old-timers—many of whom were oddball sons of the rich who took to biology and butterflies rather than the family business—

Waldo Schmitt

were often a frugal lot who stayed in rooming houses, ate off hot plates and in cafeterias, and spent all their time in the Museum working on the collections. Even if they didn't have small fortunes tucked away, they scrounged for field-trip money or paid their own way. Some of the wealthy ones, on the other hand, didn't even bother to cash their paychecks, thus throwing the accounting office into turmoil.

Again, Schmitt was a case in point.

Born in Washington, he became interested in biology at an early age when his uncle would take him and the other Schmitt boys on very prolonged Sunday hikes (*Spaziergange*, they called them), collecting and identifying insects and butterflies, pond and stream life, rocks and minerals and wildflowers.

He joined the Smithsonian in 1910 as a "temporary" assistant, an arrangement that was to last for the next 67 years. It also initiated one of the classic tales of the Natural History Museum that has almost become a fable, involving a rather homely, plain old maid who loved to study crabs and a young newlywed who loved to study shrimp. It was a romance, but a romance of science.

The story goes back to 1914 when "The Crab Lady," Miss Mary Jane Rathbun, was slaving away on a miserly salary and single-handedly running the Museum's Division of Marine Invertebrates. At age 54, and with only a high school diploma, the tiny, self-trained Miss Rathbun was a renowned carcinologist just beginning her monumental, four-volume monograph, *The Crabs of America*.

Schmitt was so competent that Miss Rathbun wanted desperately to hire him, but the parsimonious pursers in the Smithsonian "Castle" set up a squawk. Faced with this word, and fearful of the fate of her crab opus, the prim, tenacious lady, who happened to be well-to-do, decided to strike a deal. She would resign as head of the department, forfeiting her own salary, if Schmitt were hired. The bookkeepers sensed the bargain they were getting and took it.

The day after her resignation, Miss Mary was made an "honorary research associate." She continued to come to the office daily for almost the next 30 years without pay. Later, after acquiring so much knowledge in her field that it was getting embarrassing, she was awarded an honorary Ph.D. She ended her federal employment precisely the way she began it, 62 years before, with the old U.S. Fish Commission in Woods Hole, Massachusetts—as a volunteer. She died, practically at her desk, in 1943 at the age of 82.

Even as late as 1963, Schmitt, 76 and retired for six years, was back in his beloved Antarctica collecting for the Smithsonian, bringing back 29,000 specimens. As true with all the hardy old-timers, he set a pace that shocked the younger scientists, who couldn't keep up with him. He cut such a remarkable figure that "Schmitt Mesa" was

Schmitt and Mary Jane Rathbun

The Tradition

Leonhard Stejneger

named after him—a 30-mile, ice-covered series of outcrops west of the Weddell Sea at the base of the Antarctic Peninsula—a desolate region that only a man like Schmitt could love.

Schmitt had a sincere affection for Miss Mary, as a son for his mother, but his scientific super-hero in an age that turned them out with regularity was a heroic figure named Stejneger. This Old World Norwegian commanded an almost hushed respect and admiration, but he was also a sweet, lovable old man who never lost his temper and was always ready to share his knowledge with others.

Leonhard Stejneger was a wondrous fellow. "The man's breadth and depth of knowledge in more fields than he was especially interested in was astounding," wrote Schmitt, a man not easily given to writing superlatives and certainly not one who was easily impressed.

Dr. Stejneger thought of the Smithsonian as a sort of Elysium, a happy place for happy souls committed to the study of natural science and the wonders of life on this planet. He went on to become a Smithsonian legend. He was the only honorary research associate in history kept on full salary after his retirement, and he was granted life tenure as curator of biology—a move that took a Presidential order.

A man who loved to click his heels in mid-air while dancing the mazurka, Stejneger—who wore a black cape and looked like a figure out of a mystery story—had a full and flowing mustache and a white Van Dyke beard. He studied medicine and had a degree in law. He was an accomplished artist and linguist, a classicist and a botanist. He was a world authority on ornithology and herpetology, and took part in international congresses on zoology and entomology as well.

He studied plants in the Alps; fur seals in the Arctic; salamanders in Virginia; birds in Japan; poisonous snakes in North America and, just before his death, co-authored an indispensable herpetological tool entitled *The Check-List of North American Amphibians and Reptiles.*

At the age of 16 Stejneger was keeping notes on Norwegian birds, illustrating and coloring his own pen-and-ink sketches that were to show that he would have had an equally successful career as an artist. His father wanted young Leonhard to enter the family mercantile establishment, and his ailing mother wanted him to study medicine, which he did for a time. But the training in botany and zoology given as a part of every physician's education just led him further astray and into the adventurous world of travel, exploration, and collecting.

Thomas Barbour, who went on to become the director of Harvard's Museum of Comparative Zoology, as a youth of 16 visited with the curator of the U.S. National Museum's collection of reptiles and amphibians and was forever impressed: "He had that quality that erases the barrier of years and puts in its place a serene feeling of contemporaneity." Years later Stejneger's kind patience was also to impress a 14-year-old boy named Frank H. Johnson, who became a professor of biology at Princeton.

Past the age of 80, the linguist taught himself Polish in order to better translate the old records of explorers. It was one of the few European languages he didn't know well. At the age of 85 he finally completed his magnum opus: a 50-year work that culminated in his epic, 632-page life of the great explorer and naturalist, Georg Wilhelm Steller. Stejneger died in 1943 at the age of 92, struggling into the office near the end, even after getting knocked down by a car once on his way to work. He simply got up, brushed himself off, and apologized to the driver for getting in his way.

*   *   *

T. Dale Stewart, director of the Natural History Museum from 1962 to 1965, was the son of a small-town pharmacist in Pennsylvania when he came under the strong influence of his father's drugstore clerk, who dragged him along on arrowhead hunts. Later he was to come under the almost hypnotic spell of one of the most fabulous of all the Smithsonian's Grand Old Men of Science—Ales Hrdlicka, a character of the first order.

Stewart was leading a contented life clerking in the town bank when the arrowhead man, John Baer, talked his mother into shipping him off to Washington to continue his education. Baer was already under the museum spell, and soon Stewart was hooked for life. Stewart roomed at Baer's house along with a young archeologist named Henry Bascom Collins, who was eventually convinced by Hrdlicka that he should devote his life to studying Eskimos.

The thing that most interested Hrdlicka about Stewart was the youth's bookkeeping background. By God, the boy was able to operate an adding machine, the better to tabulate his endless reams of figures. Hrdlicka was known as "The Skull Doctor," and he worked with a collection of human skeletons that was more numerous than those in most town cemeteries. Working with the man who became the father of physical anthropology in America was a demanding task. There was no question of coming in at night or on the weekends if needed, and the thought of extra pay was ludicrous. Hrdlicka turned back portions of his own salary to help finance expeditions to the wildest corners of the world.

A formidable-looking Czech who prided himself on his full, shaggy mane of hair, he believed in total commitment to the study of physical anthropology, or any other subject he happened to be interested in (and there were many, including psychiatry and the insane). Largely self-taught in the field of physical anthropology, he simply could not get enough human skeletons and skulls to study although the Smithsonian had thousands. He robbed graves with an enthusiasm and fervor that would be considered quite eccentric today, and might even have gotten him shot. No one has ever collected skulls like Hrdlicka, who thought that burying them was completely irresponsible and a waste of valuable scientific material.

Curiously, Hrdlicka himself did not quite go along with his belief, even though he campaigned tirelessly for other Smithsonian scientists to leave their brains to the Institution for study. He desecrated his own skeleton instead of adding it to the 25,000-member boneyard he had built up from practically nothing to what now has become the largest collection of its kind in the world. Hrdlicka willed his cremated remains to the Department of Physical Anthropology "in perpetuity," and they sat around his department in an urn for years and years in his sort of mini-museum along with his death mask, some of his custom-made measuring instruments, and the cremated remains of the first Mrs. Hrdlicka.

Stewart followed the trails of Hrdlicka to some of the most inac-

Ales Hrdlicka, circa 1903

Department of Anthropology, 1931.
Front row, left to right: T. Dale
Stewart, Frank M. Setzler, Neil M.
Judd, Walter Hough, Ales Hrdlicka,
Herbert W. Krieger, Henry B.
Collins, Jr. Back row, left to right:
Charles T. Terry, Jr., William H.
Short, Richard A. Allen, George D.
McCoy, William H. Egberts, Richard
G. Paine, William H. Bray, Leta B.
Loos, Helen Heckler.

cessible regions of the world. It was a mighty accomplishment just to get to some of those places. No wonder so many of those early naturalists qualified as members of the Explorers Club. Hrdlicka, who slept on a board at home to be ready for sleeping on the ground in the barren, fog-bound Aleutians or in Siberia, was never one to complain. Mosquitoes were the only things that seemed to bother him. But in one moment of candor, he confessed to Stewart over his luncheon mush of crackers and buttermilk: "If you ever want to have your spirit chastised, try Siberia." His physical endurance was amazing. Between age 60 and 75 the indomitable skull man headed no less than 10 expeditions to the most Godforsaken reaches of Alaska and Russia to find evidence for his contention that America's first inhabitants came from Asia by way of the Bering Strait—a position that provoked a major scientific controversy.

Born in 1869 in what was then Bohemia, Hrdlicka immigrated to this country at the age of 13 and worked in New York at a cigar factory while studying Greek and Latin in evening classes with the Jesuits. Eventually he learned to read virtually all European languages. Studying medicine under a New York physician, Hrdlicka got his M.D. degree but became sidetracked and pulled away from the general practice of medicine while on a bone-hunting trip to Mexico. He came to the Smithsonian as its bone man in 1903 where he was given a place to sit and paper to write upon, but little else.

When he got word of a bonanza of human bones that the Smithsonian had turned over to the Army Medical Museum, Hrdlicka's lifelong search began and there was no stopping him. He demanded the return of the bones, and back they came.

Dressing like an Old World European even in the early 1940s, he wore dusty, black outfits and laboratory smocks. Around a high celluloid collar he clipped on a made-up tie that covered a pleated dickey shirt of the type hardly sold anywhere at that time. He was gruff, abrupt, and disconcerting, and he never understood women. He objected to their smoking, wearing makeup, mixing with male scientists, and working anywhere except in the home, "vere dey belong," he told Stewart.

His ideas of office behavior were curious to say the least. He forbade any kind of conversation among the staff, believing that more work got done in what amounted to solitary confinement of a sort. He once caught Stewart, whose wife had just had a baby, in the act of *laughing* together with a colleague in the office. "Vat is *DIS?*" he roared. Stewart explained and Hrdlicka relented a bit. The old man never had children of his own, but he loved them—especially if they were male. Hrdlicka asked Stewart, who always called him "Sir," the obvious question: "Vat is da sex?" Stewart lowered his eyes and confessed that it was a girl. "Oh, my," said Hrdlicka, sadly shaking his head over the terrible news. "The first one is usually a veakling." Stewart eventually brought his daughter to the office where Hrdlicka held her in his lap and felt her bones and probably measured her skull (he was forever measuring people's skulls). "But she is strong!" he said, genuinely puzzled.

Stewart, who came to the Smithsonian and to Hrdlicka in 1924, was encouraged to get his medical degree before he could ever hope of succeeding the Great Man, who retired in 1942 because of ill health and died soon after at the age of 74. Stewart succeeded Hrdlicka as a sort of non-practicing Museum resident house doctor. Medical doctor or not, Hrdlicka almost believed in a universal balm: his favorite prescription for everything from dandruff to fallen arches was citrine ointment, a huge jar of which he kept near his desk.

*   *   *

Before World War II, the Museum was a difficult place for a woman scientist to be. The male naturalists, who often didn't trust one another, were certainly not expected to trust women, were they? Even if the female of the species was knowledgeable and shared the love of beetles, for example, it was hardly reason for a man to actually turn over some of the plus-7 million beetles for her to study.

Doris Holmes Blake, an accomplished artist and research associate who studies Central American beetles, was known for years as "The Lady in the Attic." In 1978 she ranked as the undisputed Grand Old Lady among dozens of Grand Old Men who continue to work regularly without pay after their retirement. But Mrs. Blake has never retired because she was never really hired. Now 86 years old, the courtly coleopterist has worked, without pay, at the Smithsonian for more than 40 years. And for 30 of those years she was sort of exiled to the attic before her "dear" museum friend, the late Doris Cochran, curator of reptiles, found room for the beetle lady in the Division of Reptiles in the early 1960s. The "Two Dorises," as they were known, used to go off on expeditions together and once went to South America on a collecting trip.

Again, and again, and once again, it all started in childhood. "My word, I thought, who would want to play with dead dolls stuffed

T. Dale Stewart

Doris Blake

with sawdust when you could have all the live toads for the taking?" Mrs. Blake recalls. As a young girl she "began catching grasshoppers and storing them in my pinafore pockets as soon as I could toddle. Later it was frogs. When I reached adolescence I used to betake myself regularly with the grievances of sensitive youth to the New England woods and old pastures. There I would promptly forget the woes of being misunderstood by tracking down a rare fern, hunting a chewink's nest, or merely standing in pantheistic ecstasy among the great beeches near Flot Meadow." Her Bible was Thoreau's *Walden*.

Mrs. Blake met her husband on the first day of first grade. He was able to stand up and recite the alphabet. "I was so impressed, and told my mother, who that evening taught me the alphabet." The next day, young Doris got up and recited the alphabet and impressed young Sidney. The maturing naturalists went on to high school together, and the school yearbook predicted they would marry one another "and live among plants and wild animals."

Doris Holmes married Sidney Blake and began illustrating his published works. Dr. Blake, a botanist and famous taxonomist, was reading proofs of the second volume of his 749-page *Geographical Guide to Floras of the World*, when he died in 1960 at the age of 67.

Mrs. Blake became fascinated by beetles when she came to Washington in 1918 and got a job working for an old bachelor coleopterist at the Division of Truck Crops Insects at the Agriculture Department's Bureau of Entomology, always closely associated with the Smithsonian Institution.

Her boss was Frank H. Chittenden, who had an honorary doctorate but never even received his bachelor's degree from Cornell, where he spent most of his time studying what he wanted to. Working in the school lab, and collecting and mounting insects far into the night, he was largely self-taught and came to know as much about entomology as his entomology professor.

He was typical of the old-time naturalist, "born with an abnormally large collecting and hoarding proclivity," Mrs. Blake wrote. "Added to it, there is usually a total indifference to appearance. Shortly there results a disorder not equalled anywhere except in a junk shop." It was "characteristic of him that he never learned to tie even a bow knot for his shoe lacings." She once found a mouse nest on his cluttered desk.

She also came to know the Smithsonian's Eugene Amandus Schwarz, the dean of Washington entomologists, by traveling to and from Dr. Schwarz's office to borrow beetles for Chittenden to study. The two men never got along. Chittenden would hoard the beetles, swearing he never had them in the first place. Schwarz would sneak into his office at night and steal them back, sending Chittenden into a rage. Schwarz disliked Chittenden so much that he wouldn't even name a new beetle species after him—an unforgivable affront to a fellow entomologist since there is a continuous shortage of names to call new insect species being discovered.

Toward the end of their non-relationship, Chittenden did for Schwarz what Schwarz never did for him—he named a new species of beetle after him. "*Sphenophorus schwarzi*," he called it. Mrs. Blake took the specimen over to Schwarz, along with the certification papers. "Oh, I remember that thing," Schwarz said. "I remember it. I am glad to see it again. I am that." It turned out to be a beetle that Schwarz himself had collected a long time before, had foolishly loaned to Chittenden, and had never seen again until that day.

Schwarz was like so many naturalists who shocked their families by choosing to study bugs instead of obeying parental wishes. His parents had wanted him to be a professor of philology, but he rejected that idea and left his home and Germany with his beetle collection —never to return there again. And all for the love of insects. He almost died when the U.S. government tried to take away his collection when he landed at Ellis Island.

* * *

Eugene A. Schwarz and, bottom photo, Frank Chittenden.

Entomology has a particular way of attracting single-minded determination from its devotees. They know that they will never live long enough to understand a fraction about the insect that they study. A third of the species of the arthropod world have not even been collected, and insects become extinct and evolve new species before there is time to get around to finding what is, let alone what was.

Maybe it was this hideous frustration that drove at least one of the old Smithsonian bug men a bit buggy. He went underground, literally. He started digging tunnels under his home, and his "other hobby" (he said he loved the smell of fresh earth) was discovered only when a truck fell through the street. But Harrison Gray Dyar, a cantankerous and wealthy man, also had secret tunnels under a second Washington house. This is where Dr. Dyar maintained his secret second family, for Dyar was a bigamist as well.

Those Smithsonian people who take themselves too seriously object to digging up stories about colleagues like Digger Dyar. "It makes us all look like crazy people," they complain. Well, there's no getting around the fact that what many of them have chosen to do for the rest of their lives is a little bit curious to the average outsider who, sad as it seems, is not driven by anything other than the search for financial security. Only those who collect things (and there are hundreds of thousands of this breed) can understand what drives an obsessed naturalist. The collection and the systematic notes accompanying it are what add up to knowledge about the collection and what it means to all of us. The collector/naturalist can never get enough of what he collects. Never! He lives in fear of someone who has more, or better, or has discovered something that he is trying to discover.

If you can't locate the fossil, then find it another way. In one extreme case, an Australian dragonfly expert made regular trips to Boston, where a medium put him in a trance and "transported" him back to the Paleozoic Age so he could have a look-see. The Australian then described the giant prehistoric dragonflies in scientific publications. Only years after his death were the first dragonfly fossils from that period found. His descriptions were said to be remarkably accurate.

Dyar, who died in 1929, inherited a fortune and was not paid a cent for 31 years while working for the Agriculture Department and the Smithsonian. He traveled all over the world at his own expense and "on the basis of his knowledge of adults, larvae, and eggs," says J. F. Gates Clarke, "his work may perhaps be considered the basis of our modern classification of moths." Dr. Clarke says Dyar's greatest

William Schaus

work, however, was his four-volume *The Mosquitoes of North and Central America and the West Indies.*

Sarcastic and extremely critical, Dyar took great pleasure in ripping apart colleagues in scientific print, thus enlivening the dull journals of his day with his unpredictability. The scientist's word was considered sacred, and everything was published exactly as he had written it. One of his continuing feuds was with an obese bug man who was very sensitive about his weight. The insensitive Dyar, however, named an insect "corpulentis" after him. The fat man, in turn, named a particularly loathsome-looking bug "dyaria" after Dyar, who was as disliked at the Smithsonian as his colleague, William Schaus, was liked.

For every true eccentric like Dyar, there are dozens more like Dr. Schaus, another rich boy who infuriated his parents by showing no interest in the family art dealership in New York and taking flight on the wings of lepidoptera. No matter that he was schooled in Europe in music, art and language; when he was a young man he came under the influence of an entomologist, and his fate was sealed.

The bachelor Schaus was one of the great contributors and dedicated workers in the lepidoptera collections, says Clarke. Coming to the Museum in 1895, he retired in 1938. He described 5,000 species. He took his work home with him every night in the form of a small box containing six specimens, each representing a different species. In the morning he would return with each species described and each specimen labeled "type."

Schaus traveled all over the place with his English valet and companion, Jack Barnes, and collected more than 200,000 lepidoptera. Yet most of his vast collection was never counted. In 1909 he sent a typical message to the Smithsonian, this time from Costa Rica:

"I announce to you the gift of my butterflies and *Sphingidae* to the Museum, as they were not included in the large collection of moths I gave the Museum three years ago—I am glad to be able to do so—I am still hard at work and securing many new and rare species, so there is no danger of the Smithsonian losing its foremost place as possessor of the finest collection of Tropical American Lepidoptera."

Barnes became Schaus's life-long companion in his tireless search for butterflies. It is not recorded if Barnes shared his employer's enthusiasm for the chase with the butterfly net, but he certainly shared in other ways. Early in their friendship, Schaus bought his friend a quantity of stocks which appreciated in the crash of 1929 that knocked Schaus for a loop. As it turned out, the valet became wealthier than the master and was fixed for life when the master also left him a valuable stamp collection after his death in 1942.

*   *   *

Agnes Chase was the legendary figure in Botany, a Museum department that is legendary for legendary figures. Self-taught in the field that she absolutely mastered, she became the dean of American agrostologists and came to know more about grass than any person alive. When meeting a stranger, her first question was often, "And what grasses do *you* work on?" If the person was not into grass, she turned on her heels and walked away. "Grass is what holds the earth together," she wrote. "Grass made it possible for the human race to abandon the cave life and follow herds. Civilization was based on grass, everywhere in the world."

As custodian of the largest collection of grasses in the world, shortly before her death at the age of 95 she published her three-volume index to the grasses of the United States with more than 80,000 entries. Although retired in 1939, she continued to come to work for the next 25 years six days a week, for free. "She was almost in a state of anguish when she had to stay home," said a friend. "She would plead with her colleagues to take her to the office, although she was scarcely able to walk."

She finally died in a nursing home after being there for one day.

"Grass fascinated me when I was a little girl," she once said in an extremely rare interview. "I used to make grass flower bouquets for my grandmother." Pressed for more, she added: "I simply don't have time to talk to you. I'm too busy. I can't talk with you. There's not much time left."

Twice she was jailed in Washington for her suffragette activities, and once in South America where she was locked up as a suspected lunatic. The aging lady, who would go on lone expeditions into the wildest interiors of Brazil in search of grass, was observed crawling on her hands and knees and pulling up clumps of grass. The natives thought she was starving and forced to eat grass.

She once took an extension course from the University of Chicago, but she never received a college degree—until at last at the

Agnes Chase

The Tradition

Agnes Chase with her pet squirrel, Toodles, around 1914. Opposite, ethnological collections being sorted, circa 1890.

age of 89 when the University of Illinois named her an honorary doctor of science. She died in 1963 and was buried in Chicago. She left no survivors except the grasses she was known by.

*   *   *

Behind every one of the "magnificent foragers" at the Museum—be they teen-age volunteers like Jason Weintraub or middle-aged men like Donald Davis—there is an older one whose spirit, and often his ghost, lingers in the collections. There is the specimen collected by him, named by him, mounted by him, labeled by him, described by him.

And so it goes: this "increase and diffusion of knowledge among men."

It's a wonderful line from the obscure Englishman named James Smithson—the "antenuptial son of the first Duke and Duchess of Northumberland"—and it's been the reason and paid the way for all these years, for all these people who don't accept pay, for all these gentle souls who do it out of pure and honest love of scientific knowledge.

To get the beetle; to find the bird; to dig the fossil; to live the museum life . . . happily together, forever after, in the house that Smithson built.

II. The Solitary Researcher

Biology is often a singular, individualistic pursuit: a lone scientific intelligence, occasionally blessed with a little help, out in the field—observing, experimenting, learning.

Lights To Hide By

Off Oahu, Hawaii

Working in pitch darkness, invertebrate zoologist Clyde Roper slips his hand down into the ice-cold water of an aquarium tank to prod a small squid and see how it will react. Doing so, he brushes his hand against one of its arms. Reacting violently, the squid rushes at his hand, its arms flailing. Simultaneously there is a brilliant flash of light, like a camera strobe going off, startling the onlooking scientists.

The squid has activated a set of bioluminescent organs called photophores. These are located on its armtips, as they are in some other kinds of oceanic squid, and Dr. Roper and his co-worker, Dr. Richard Young, are the first to measure this flashing phenomenon—believed to serve the animal as a defensive warning. "I know that if I were a fish and a squid flashed at me," says Roper, "I wouldn't stay around to try to eat him. I'd be gone in an instant."

Bioluminescent animals are found everywhere in the open ocean, but little is known about the role that this lighting capability plays in their lives. Roper studies the mechanism in cephalopods—a group that includes squids, cuttlefishes, and octopuses. His interest in their bioluminescence was first whetted as a graduate student when he caught an oceanic squid previously not known to be luminescent. To his astonishment, it lit up. Ever since, he has wanted to learn more about how the luminescent organs function to help the animals survive.

In the oceanic depths the most common cephalopods are the small- and medium-sized squid. Most of the photophores on the body are on the ventral sides of head, arms, eyes, and viscera. Many fishes

The University of Hawaii's research vessel, Kana Keoki, *aboard which studies of bioluminescence in deep-sea animals were conducted.*

The Solitary Researcher

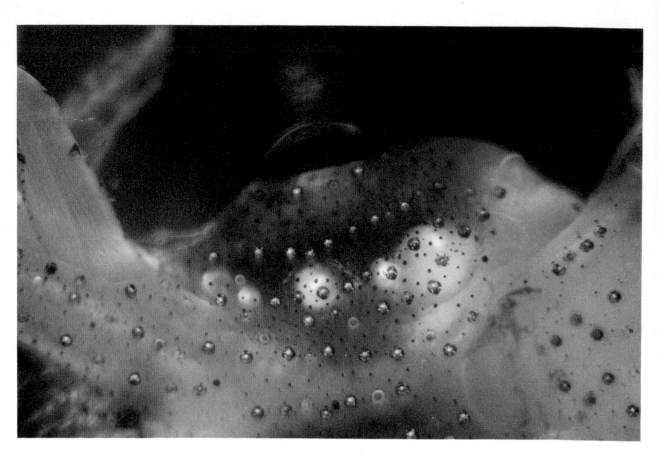

Ventral view of one side of the head and funnel of Abralia trigonura, *a small midwater squid (total length 2 inches), showing the numerous tiny photophores used for countershading. The large, pearly organs beneath the skin counter-illuminate the completely opaque eye, top, while the largest one is used as a flasher to startle attackers. Opposite, the rarely caught squid,* Taningia danae, *provided the first observations ever made on living specimens of this species. The dark, elliptical bulbs on the tips of the shortest arms are large photophores that flash when the animal is disturbed.*

and crustaceans have similar patterns of photophores. For a long while no one knew what these photophores were. After they were finally recognized as light-producing organs, a German ichthyologist suggested that their purpose was to camouflage the fish against the dim daylight remaining in the mid-depths of the sea.

The idea was that the opaque form of animals at depths of 400-800 meters (daylight penetrates the water as far down as about 800 meters) would be silhouetted against the down-welling light, and could be easily seen from below. If this were true, the animal would likely end up as lunch for one of its predators. But if it perceived the quality and quantity of the dim light around it, and then glowed dimly to match the background light, it would be able to obliterate its silhouette and thus render itself invisible.

This defensive tactic, called "countershading," was considered theoretically plausible by scientists. Yet no one was able to verify it experimentally because of the difficulty of observing the behavior of mid-water animals either in the open sea or in aquariums. When caught in nets or traps and transferred into tanks on the surface, such animals tended to live for only a short time.

Intent upon demonstrating countershading in squids, Roper and Young, who is a biological oceanographer at the University of Hawaii, constructed a portable shipboard system designed to keep deep-sea squid alive for study. This involved building a dark aquarium room and an adjoining laboratory. Live, healthy squid were brought to the dark room and placed in an aquarium that lay inside a black, light-shielded box. A rheostatically controlled, downward-directed lamp shone from

The experimental tank used during countershading studies. The glowing ventral surfaces of squid are viewed in the mirror set at a 45-degree angle beneath the aquarium. The lamp, blue filter, and diffusers are located above.

above the aquarium. Through an aperture on the side of the black box, the squid's silhouette could be observed via a tilted mirror set beneath the glass-bottomed aquarium. An adjacent aquarium was used as a holding tank for the animals awaiting observation.

The squid themselves were kept as comfortable as possible by a refrigeration unit that regulated the temperature of the sea water in both the holding and observation aquariums so that it would coincide with that experienced by squid during their daily migrations from depths of 500-750 meters during the day (40-45 degrees F.) up to warmer depths of 150-300 meters at night (55-60 degrees F.).

As another precaution, the animals were collected from the ocean's depths in a special trawling bucket designed to minimize internal water turbulence and to eliminate exposure to sunlight, which might damage the squids' eyes and prevent them from responding properly during the experiments. After the trawling bucket was hauled out of the water, it was wrapped in black plastic and taken to the dark room where the catch was quickly transferred to screened vials and placed in the holding tank.

The first cruise with the newly constructed system off Oahu, Hawaii, resulted in little more than equipment breakdowns. On the second 10-day cruise, however, everything functioned smoothly. The scientists found that captured squid stayed alive in the aquariums for up to a week.

"Being able to keep deep-sea animals alive was in itself a real breakthrough," remembers Roper. "But, unfortunately, though the animals appeared to be healthy, they wouldn't light up. I'd go into the aquarium room and adapt my eyes to the dark for 45 minutes, then spend all day watching the squid images in the mirror. Young sat in the adjoining laboratory operating the experimental lamp and other equipment, communicating with me through a speaking-tube. For days we'd put new squid in the aquarium and turn on the overhead lamp. Nothing happened. For some reason the animals didn't respond. The only explanation I have is that we weren't handling them gently enough as we brought them aboard.

"Finally, having improved our boarding technique, one day I was watching two squid that were both clearly silhouetted against the dim overhead light. All of sudden, I couldn't see one very well. Then I saw it turn and swim to the center of the aquarium and completely disappear. I asked for the lamp to be turned off and there it was, glowing like a Christmas tree! You could see little jewels of light all over its undersides. It was a very exciting moment."

After this, Roper and Young found that their rate of success increased. They placed an animal in the tank and let it settle down in the dark. Then they would turn the experimental lamp on and off for designated periods. It usually worked as they hoped. When they turned on the stimulus lamp, the squid turned its lights on. When they turned it off, the squid turned its off.

On their next two cruises the scientists ran tests to see if the squid could regulate their luminescent intensity. This was a critical experiment, for if the squid actually protected themselves by countershading, they would have to be able to match the changing light conditions that exist in their natural environment over a depth range of about 500 meters.

A transparent juvenile cranchiid squid must countershade its opaque eyes with photophores (mass of yellow below eye). The liver and ink sac cast a minimum of silhouette when situated vertically, but in adults they enlarge and lie horizontally, their silhouette countershaded by a large complex photophore not visible here.

For this work an extremely sensitive photo-multiplier tube was used to measure the intensity of overhead light. The light could be set at a specific brightness and read on a meter by the scientist in the lab adjoining the aquarium room. Each squid—confined in a cage in the tank so that it couldn't swim out of the overhead light—was exposed step by step up through a series of light-intensity levels as the observer in the aquarium room watched to see if it could match each new level and become invisible.

"Amazingly, they would," reports Roper. "When we got them up through four steps we'd let them rest for a while and then continue on. We'd finally get to a level that was too bright for them to match. Then they'd show signs of stress and become physically agitated. We then reversed the process and reduced the light to find the dimmest level at which their luminosity functioned. Each species, it turns out, has a specific range of light intensities through which it can regulate. Now we can demonstrate that this corresponds to the light intensities experienced over its normal vertical distribution in the ocean."

When Roper and Young had amassed data on a variety of squid species, they wondered if they could prove that other midwater animals also countershaded. Unexpectedly, they had the opportunity to do so. Late in the cruise, a graduate assistant burst into the lab to announce that a black anglerfish—a midwater ocean fish rarely caught alive—had been brought aboard. Observing it as it swam around in a tub, Roper suddenly noticed that it glowed when viewed from the tail-end, but not when seen head-on. Immediately it was placed in the observation aquarium.

"It turned headupward," relates Roper. "At first we thought it

The sleek, black stomiatoid fishes counter-illuminate their silhouettes against the down-welling light with tiny round photophores resembling running lights. Large photophore beneath the eye flashes brightly, but its behavioral significance is unknown.

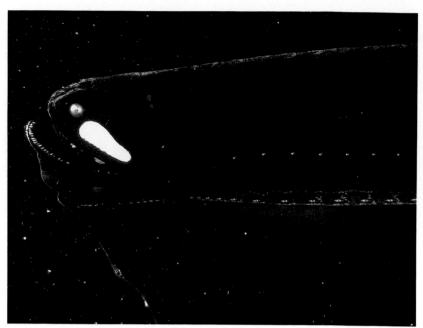

Deep-sea squids of the genus His-tioteuthis *have the left eye greatly enlarged. Though it has been suggested that the huge eye is designed to gather very dim light in the great depths, its actual function is not known.*

was sick, but we soon realized that this was how it oriented itself in the deep sea. That's why its countershading photophores were visible from the tail-end. We put the fish through our light-intensity regimen and it countershaded at every step until we finally surpassed its ability to respond. At this bright level, it dove toward the bottom of the tank trying to get down to a deeper, darker level.

"Next we ran a species of half-red shrimp through the tests. It had many of its photophores on its legs, and we wondered how they could effectively countershade the animal. It turned out that the animal spends much of its time in a quiescent state with its legs tucked up under its body, a position that enables it to countershade very well."

Almost all of the deep-sea animals that Roper and Young studied had the capability to use bioluminescence to do more than countershade. Some were observed to expel glowing clouds of luminous material instead of dark, opaque ink like their shallow-water relatives. Almost all of the animals had special pairs of light organs that were used as flashers to warn away potential attackers.

Many more bioluminescence experiments are planned for the future by the two scientists. One matter they want to explore is the degree to which the eyes of the squid are responsible for sensing the light and controlling the animal's light output, in comparison to the special photosensitive vesicles under the skin that also act as light meters.

"As we go along we're making considerable progress in understanding bioluminescence in cephalopods, and, since bioluminescence is a nearly universal characteristic of deep-sea animals in general, the results have significance for all of deep-sea biology," explained Roper back at the Smithsonian. "Plus, there are few things in nature so beautiful as these canny, functional, biological lights. They lend a real aesthetic dimension to our work."

Pyroteuthis addolux *countershades its eyes and viscera with large light organs that reflect the camera's strobe flash. The two photophores that underlie the liver (lower part of picture) can be rotated so the liver will be countershaded against overhead light regardless of the squid's orientation.*

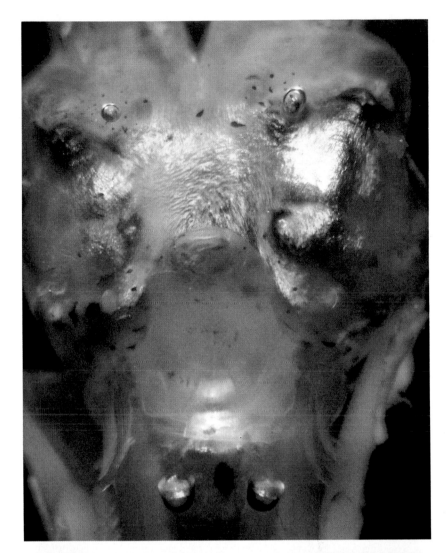

Photophores of Histioteuthis *are directed obliquely forward, presumably to countershade the animal as it hangs at an oblique angle in the deep sea. Although a number of specimens have survived for days in the aquarium, they have not yet been induced to luminesce there.*

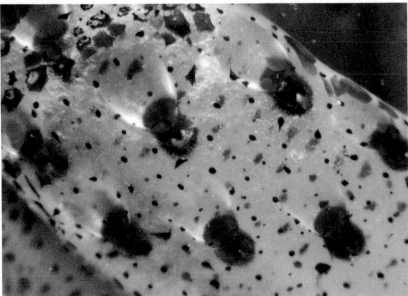

Migration in the Himalayas

Bhutan

Every spring, northbound migratory waves of Palearctic birds cross the highest passes of the Himalayas on their way to breeding grounds in Siberia. The birds fly through again, southbound, in August and September to winter on the Indian subcontinent and in Southeast Asia. But this invading horde of transients does more than merely fly by. It feeds prodigiously en route, consuming a large part of what little food supply is available in the high-altitude Himalayan zone, inevitably leaving local birds on short rations most of the year.

According to S. Dillon Ripley, an ornithologist who is also Secretary of the Smithsonian, when the Siberian-breeding species traverse the passes of Bhutan, the 35 to 38 often related species known to be summer residents in the 9- to 16,000-foot zone virtually "cling to life by their toenails." On three expeditions to Bhutan, he has observed that they make little effort to compete, and in fact are very unobtrusive, some even becoming extremely recessive in their behavior.

Mr. Ripley believes that too much seasonal competition from their migrating relatives is responsible for the low reproduction rates of birds in the high montane zone. A kind of competitive exclusion results among certain local species, which will probably lead to some of them becoming extinct and being replaced by others. Evidence of great environmental pressure on the local birds is provided by the fact that pairs of some species rear only two instead of the normal clutch of four to six young.

Looking northwest toward the eastern peaks of Chomo Lhari, which rise above 20,000 feet. Baggage yaks are visible on the horizon at 15,000 feet.

The Magnificent Foragers

Blanford's rose finch, above, and a redstart, right. Indigenous to the high country, both are subject to fierce competition from migrating species.

A.R. Tangerini

The Rumjum, officer of a district near the Tibetan border. Expedition tents can be seen in the left background.

Surrounded by asters, marigolds, wild carrots, primulas, and edelweiss, a baggage yak, used for transport above 10,500 feet, stands in an alpine meadow in northwest Bhutan.

As an ecologist, Ripley is particularly interested in the Himalayan peaks because they form the only significant chain in the world running east and west across a major bird migration route, thus serving as a protective barrier between the arid Palearctic zone to the north and the warm, lush wintering grounds to the south. Though, in one sense, this enormous 1800-mile ecotone functions as a temporary, island-like refuge for migratory birds, they are also subjected to incredible stress as they traverse it. They must climb to higher altitudes, and thus face more severe climatic extremes, than birds on any other major migration circuit.

During Ripley's 1973 expedition an unseasonably late monsoon caused heavy mortality among migrants. In one transect at 9,000 feet he observed a great number of semi-mummified birds that had perished from snowfall at 14,000 feet, considerably more than would have been devoured in a short period by predators. An accompanying migration of hawks, buzzards, and kites constantly preys on tired and weakened birds.

The study of high-altitude populations and zoographic distribution of birds along the Himalayas presents inordinately difficult problems, according to Ripley. Enormous effort is required in the field to overcome the altitude, cold, wetness, and darkness of the region, all of which hinder the systematic observation required to produce precise data. If Ripley resumes his research in Bhutan, scientific curiosity will be only part of the attraction. For, in his words, the high peaks of the Himalayas will always offer a thrill unmatched by any other experience.

The Magnificent Foragers

Above, Mr. Ripley during a stop near Bumthang, a valley about 9,000 feet above sea level. Right, while some nations contemplate space shuttles, the Bhutanese are still grappling with getting their primary roads built. Beasts of burden such as the yak are often more reliable transportation than cars, and certainly better able to move through mud, as Mr. Ripley and his party discovered.

The Solitary Researcher

Thwapping Skippers

Georgia-Florida border

Though one hundred and sixty-five thousand different species of moths and butterflies have been cataloged, there may well be almost as many more still undiscovered. Because of the great biological and economic importance of these showy insects, Smithsonian entomologists seek not just the species that have eluded description and classification, but also those already "described" that are not really adequately known or understood.

John M. Burns, for example, combs northern Florida and southern Georgia for certain kinds of dusky wing skippers: small, fast-moving, velvety, blackish-brown and gray butterflies that fly for only a few short weeks in early spring—and then, only when the weather is warm and sunny.

After driving back roads through monotonous, piney flatlands for hours to find the mixed oak, huckleberry, and saw palmetto scrub that these skippers prefer, Dr. Burns abruptly pulls onto a sandy shoulder beside a thicket, almost certain that his butterflies will be there.

"Before wading into the scrub," he relates, "I goop up my legs to repel chiggers and ticks, sling on an old army sack full of cyanide jars and vials, and grab my collecting net. Its long, slender handle of leather-wrapped hardwood is the shaft of a warped golf driver I bought from the Salvation Army for 50 cents; its small, round hoop is hand-shaped spring steel; and its tapered, gray Dacron net bag is a labor of love, sewn by my wife. This net has been banged, ripped, patched, bled on, and generally abused beyond belief for two decades," he declares. "But by now it's a natural extension of my arm."

Nothing, according to Burns, is more distorted than the public image of a lepidopterist dashing pell-mell through field and forest wildly waving a net. He goes after dusky wing skippers slowly and patiently. "Dusky wings often rest on the ground where I get them by 'thwapping,'" he explains. "That is, I pull up on the tip of the net bag with one hand, push down on the handle with the other, slowly bring the cocked hoop in a horizontal position over the quarry (being careful throughout to keep all shadows off it), and then suddenly let go of the bag. The hoop slaps flat on the ground with a sharp 'thwap' and the skipper rises in the bag, where I bottle it."

However, in Florida scrub, dusky wings don't often stop. So Burns has to anticipate their erratic flight path and try to take them with an ultrafast swing as they scoot by. This tactic has worked well: by his own count, his Florida batting average is about .770.

"I've been trying for 20 years to understand the evolutionary processes by which populations diverge genetically and single species give rise to two or more daughter species," says Burns. "North Florida and south Georgia are particularly interesting since they form a contact zone between two subtly differing, but very closely related, populations of dusky wings. The northern one ranges over eastern North America, from southern Canada to the Gulf of Mexico, and the southern one occupies peninsular Florida. Are they different species, or subspecies, or what? I want to find out whether there is a gap between these geographically complementary forms, or whether they meet each other or overlap; and, if there is contact or overlap, whether they interbreed and, if so, to what extent. This analysis will not only throw light on evolutionary mechanisms but also promote the best possible classification of the populations in question."

At the Museum Burns turns over his catch, already pinned in the field, to Marc Roth, a Department of Entomology technician. Roth sticks the skippers in a moist box to relax their muscles. Then he puts them on spreading boards, where he positions their wings meticulously, keeping them flat with taut paper strips until dry. On the pin below each specimen he puts a small printed label telling exactly where and when, and by whom, it was caught. Thus prepared, a few hundred critical skippers come back to Burns in glass-topped drawers so he can measure wings, evaluate color and pattern, and dissect and compare genitalia and various other anatomical features.

But if Burns can't find what he wants by these comparisons, he resorts to a more exotic system. To the horror of some of his colleagues, Burns likes to say, "When in doubt, homogenize!"

After returning to Florida and Georgia for more dusky wings, he deep-freezes them in dry ice to keep their proteins chemically active. Back at his Museum laboratory, he homogenizes each skipper and puts the homogenates separately in an electric field. This causes protein molecules (which carry a surface charge) to migrate. Whether a protein moves toward the positive or the negative pole and how fast it moves depend, respectively, on the sign and the size of its charge. With this technique of electrophoretic separation, he can detect not only differences among various kinds of proteins but also—more important for his studies—variations in the same kind of protein from one individual skipper to another.

The moth-like wing below is actually that of a male butterfly, Erynnis telemachus, *a dusky wing skipper first described in 1960 by Smithsonian entomologist John Burns. Wing scales in the male resemble fur, though what function the "furriness" serves is unclear. Females by contrast have shingle-like scales.*

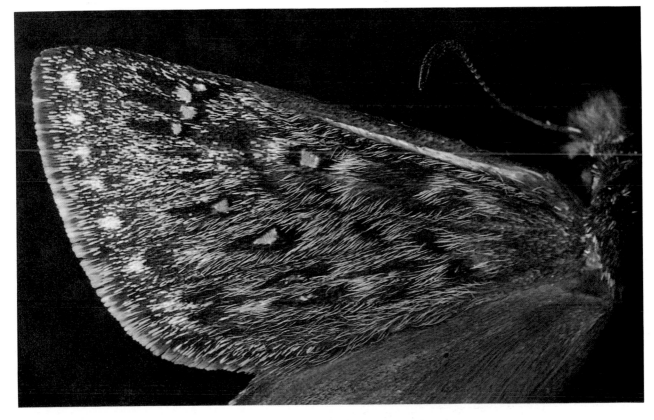

In search of cave moths, above. It is thought that caves may have been the first haunt of the moths that feed on fur coats and sweaters. Right, a hairstreak butterfly. Elaboration of its tail to resemble antennae may be a protective device to confuse avian predators, since the hairstreak appears to be flying in the "wrong" direction.

The Magnificent Foragers

A moth in wasp's clothing, Carmenta welchelorum *not only looks like a wasp but flies like one, mimicking the characteristic darting and hovering flight path of the wasp.*

"Even species that are superficially the same, or nearly so, have a considerable number of genetic differences," he states. "These are expressed as protein differences, which electrophoresis will often reveal. It is, therefore, a powerful tool for telling similar species apart. I began using electrophoresis in 1966, when it was almost unknown in evolutionary and systematic biology, to make sense of extraordinarily confusing populations."

All of the Smithsonian's moth and butterfly specialists carry on a variety of collecting activities. William D. Field, for example, pursues the hairstreak—blue and copper butterflies of North America. Donald R. Davis has collected moths in Hawaii, Mexico, the West Indies, the Philippines, and Sri Lanka. In recent years he has become fascinated by cave moths, which have led him to become a spelunker, exploring caves throughout the world. The cave environment interests him because it may have been in caves—an age-old haunt of bears and other furred animals—that insects like the first clothes moth evolved.

J. F. Gates Clarke, honorary research associate and the Museum's senior entomologist, has spent more than two decades collecting the moths of the Pacific. Dr. Clarke is intrigued by the question of how the moths succeeded in colonizing islands that are thousands of miles apart. "Moths fly only short distances," he says, "unlike birds or even butterflies. It is possible that insects may have been picked up in cyclonic disturbances and distributed, but there are still unanswered questions. For instance, it's been estimated that the odds are 10,000 to one against a pregnant insect being caught in such winds and surviving. One thing is clear: if moths make such trips, they could not be the fragile little creatures they are often thought to be."

W. Donald Duckworth studies clearwing moths, which in their caterpillar stage bore into and damage vegetables, shrubs, and trees. One of these, the peachtree borer clearwing, was very difficult to collect until recently. Like most clearwings, it does not fly at night and thus cannot be caught in light traps, the normal way to collect moths. It is also a mimic.

"Clearwings don't look like moths, and they don't act like moths," Dr. Duckworth points out. "Over millions of years they've evolved radical structural and behavioral modifications that serve as defenses against their enemies. They resemble bees or wasps. Their bodies are banded in yellow or orange, and their wings are narrow. They fly in an erratic pattern, tend to buzz, and are terrifically fast. I don't know how many times I've gone out into the field looking for clearwings and haven't been able to see any—much less collect them—even though I figured they had to be around."

But recently chemists have succeeded in artificially producing the pheromones—chemical sex attractants—that female peachtree borer clearwings emit to lure the male. The chemical can be used in some cases to reduce the numbers of male peachtree clearwings and control the damage they do.

"For reasons we don't understand yet," adds Duckworth, "the pheromones tend to attract other kinds of clearwings, too. Hundreds of different clearwing species have been trapped as a result, including many that we were never aware of before. This has been a boon to my studies. Our knowledge is often greatly advanced when new research techniques like this reach us in the field or in the laboratory."

Chub Nests

Central Appalachia

On a May morning, the Smithsonian's Ernest A. Lachner sits in a chair beside a swift Virginia stream. Through binoculars clamped to a tripod he observes a spot in midstream where a school of tiny fish hovers just beneath the surface of the clear water. Occasionally he takes a small tape recorder from his pocket and dictates his observations into it. He has been out since dawn and plans to camp here until the next day, beaming his flashlight out on the water now and then to keep an eye on the fish at night.

For more than 10 years Dr. Lachner has spent such spring weekends in the field, researching the spawning behavior of minnows and chubs. Now he finds what he is looking for—a male chub preparing to spawn. The chub selects a site and, by nudging or carrying stones in its mouth, clears the rubble away, creating a pit often a foot across and a foot deep. Then, selecting small, more uniform pebbles, it reverses the process and deposits these in the pit as it builds up a nesting platform. Various smaller minnow species are attracted to the nest site and cluster over it for their own spawning, an intrusion the chub does not seem to mind.

"This intimate attraction may explain why many different hybrid minnows are found in streams inhabited by chubs," notes Lachner. "The congested and mixed-up minnow traffic over the chub nest causes cross-fertilization."

As the hour of spawning draws near, the male chub becomes highly emotional. Its colors deepen and its body and fins flex, behavior designed to attract females. It approaches the nest from below

At upper left of photo, a male blue-head chub, Nocomis leptocephalus, *brings another stone to the nesting platform it is constructing. The smaller fishes nearby are rosefin shiners,* Notropis ardens.

A male crescent shiner, Notropis cerasinus, in full breeding dress, top. Males develop bright colors during spawning and while defending territory over their nests. A male striped shiner, Notropis chrysocephalus, displays reproductive colors, center. It was captured over the nest of a redtail chub, Nocomis effusus. Ernest Lachner, bottom, spreads a plastic sheet on the water to smooth the river's surface so that underwater photographs of spawning fishes can be shot from the bank.

and when a female enters, the male lines up laterally to her body. Following momentary contact, eggs are released and fertilized. The female then leaves the nest as the eggs settle onto the gravel in the pit. Over the next several hours, other females will visit the pit again and again until all of their eggs are deposited.

After that, the male's motions and colors become subdued and it shows no more interest in spawning. Its next step is to protect the fertilized eggs by heaping stones on top of the gravel platform. Before it is finished, the chub may have collected as many as 10,000 stones, enlarging the platform until it becomes a mound 3 feet in diameter and 1 foot high—all this work for a fish a mere 6 to 9 inches long, with a weight of about one-quarter pound when grown.

During the original construction of the nest and for several days after it has been covered up, the male chub guards it against trespassing male chubs, striking violently at them to drive them off. At length it abandons the nest, leaving the eggs to hatch and the young to work their way out through interstices in the stones.

Since the reproductive period for each male chub lasts only a few days, Lachner's problem is being there at the right time and with the right weather. "Because it rains so much in April and May, you've got one strike against you even before you go out," he acknowledges. "But if you don't know exactly where they build their nests, and in what kind of stream, and exactly what fishes to look for, you've got a lot of strikes against you. That's why there's very little knowledge of backyard North American freshwater fish—one out of every 10 species doesn't even have a name yet."

When the reproductive phase is over and Lachner has completed his behavioral observations, he abandons binoculars for a net or hook and line, not to catch his supper but to capture important specimens of the stream's fish for the Smithsonian's study collections.

"Many of the more desirable species of freshwater fish are being driven out of our major rivers and lakes," he allows. "You can see it at the Museum. Our collections from the 19th century reveal that the fish of some waterways at that time—sensitive species that were adapted to clear water—are gone. They've been replaced by fish that adapted to muddy or turbid waters. One example is a little Ohio catfish called the scioto madtom, described only a few years ago by my colleague, Ralph Taylor, an authority on American catfish. Now we think it may be extinct.

"We had marvelous freshwater fish in the Great Lakes—lake trout, ciscoes, whitefish, and smallmouth bass—and we've polluted these waters heavily," he says, slowly putting away his gear. "Certain large river species have managed to survive in small streams and lakes, but not in the major rivers and the Great Lakes, where we've built our factories. We are also putting our native fish populations in jeopardy by introducing competitive foreign fish. In Florida, for instance, you can find all kinds of tropical fish established in the waterways—even piranha. There's no point in having these foreign fish when we don't know what their long-range impact is going to be. Let's find out something about their natural history and their reactions to other forms of life first, and then—if they're judged desirable—maybe we can think about introducing them."

He gets in his car, starts it up, and heads back to Washington.

Flightless Rails

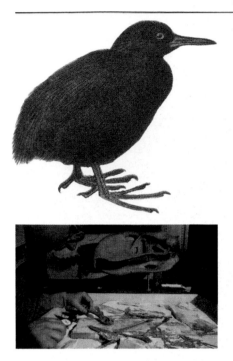

Based on skeleton studies and journal entries by explorer Peter Mundy, the extinct land-bound rail probably looked like the above drawing. Below, Storrs Olson examines the partially assembled skeleton of an extinct bird.

Ascension Island

Small, remote, and barren, Ascension Island appears as a mere dot on the map in the mid-South Atlantic. British traveler Peter Mundy stopped on Ascension several times in the 17th century while sailing from England to India. "The most desolate [land] that ever my eies beeheld," he complained. Nonetheless, he observed with interest that "a strange kind of fowle" that had vestigial wings and could not fly lived on the island. How could it be, Mundy asked in his journal, that a bird that could neither fly nor swim could be native to a tiny island in the middle of the ocean? "The question is how they should bee generated, whither created there from the beginning . . . or whither the nature of the earth and climate have alltred the shape and nature of some other fowle into this."

Charles Darwin puzzled over such biological questions when he stopped at Ascension Island during the famous world voyage of the *Beagle.* However, by Darwin's time, the mid-1800s, Mundy's strange bird had vanished, probably exterminated by rats and cats that had come ashore from vessels.

The flightless Ascension bird remained a phantom until the arrival on the island of the Smithsonian's Storrs Olson, who had decided to follow up on a British expedition report that a skull had been found on the island that might possibly be that of Mundy's bird. But, before anybody could be certain that the report was true, the rest of the skeleton would have to be located.

Dr. Olson, accompanied by a U.S. Air Force officer who had volunteered to help, drove to the vicinity of the volcanic cone where the bird skull had been found. The two men set off on foot across the loose-cindered, craggy terrain. "After several hours of wandering around poking into crevices and crawling into tunnels hung with jagged 'lavacicles,' we came upon two fumaroles," recalls Olson. "The two low cones looked like gobs of mud just erupted from the earth and solidified shortly before our arrival, though we knew no volcanic activity had been reported on Ascension since its discovery.

"The next morning we returned to the fumaroles. After much discussion we secured a rope around a mound of lava and went down, a shower of clinkers raining down around us. At the bottom we saw a dim chamber with a passageway which we followed into another chamber. Through the roof of this second chamber we could see an alternative opening to the fumarole which, we decided, would be our entrance on subsequent visits.

"On the far side of the second chamber, near the roof, we could see a narrow crevice. The officer climbed up and went through, reporting that at the end of the crevice was a drop of 10 or 12 feet into yet another chamber. Using an old frayed rope found lying nearby, perhaps left by the British expedition 12 years before, he let himself down. I lowered a flashlight and a cotton-filled box to him on a string, the latter in case he should find some specimens. Before long he called out that he had found some small bones in one corner of the chamber and, packing them in the box, he carefully sent a few up."

There in the box lay the bones of a rail, mostly leg bones but a proportionately very small single wing bone as well. A second load

Dr. Olson descends fumarole in
search of remains of Mundy's rail.

contained a few more bones, among which were portions of two
breastbones, both quite flat and definitely showing that the diminutive
bird had been totally flightless. There was no question in Olson's mind
but that he had located the remains of Mundy's rail.

"Three days later we returned to the fumarole and climbed down
the narrower but more direct entrance. We went through the crevice
and down the frayed rope into the last chamber. In a depression in
one corner of the vault, hundreds of small bones lay scattered about.
As it turned out, at least 32 birds had died in this graveyard of flightless rails. Doubtless they had fallen into the fumarole and, being un-
able to fly out, had wandered into the farthest recesses where they
died. I made two more trips into the fumarole," Olson recalls. "The
last one included Douglas Rogers, curator of the Ascension Historical
Society Museum. It was a notable day, for we recovered an intact
skull. I now had all the elements of the skeleton—more than enough
for a formal scientific description of the species."

After studying the bones at the Smithsonian, Olson concluded
that the bird was indeed a new species of rail whose ancestors crossed a
thousand miles of open ocean to colonize the island sometime in the
past. The birds encountered no predators but found the food resources
limited on the desolate island. Through a gradual reduction in the
energy available to the breast muscles, they lost their powers of flight
and evolved into the earthbound rails reported by Mundy. For these
birds, landing on Ascension signified the beginning of the end.

Beached Mammals

Atlantic Seaboard

James Mead estimates that over the past 10 years he has recovered or observed at sea as many as 50 of the 75 to 80 different whales thought to exist throughout the world. As the Smithsonian's scientific beachcomber, he specializes in the recovery of beached whales and other fish-like marine mammals.

On one occasion, Dr. Mead arrived at Beach Haven, New Jersey, in time to study a live—although critically ill—Blainville's beaked whale. It had washed ashore and then had been hoisted into a private swimming pool. Before it died three days later, Mead and other scientists were able to learn a great deal about the animal, including a possible clue as to why it is so rarely observed in deep waters.

"We were able to see how his blowhole worked," recalls Mead. "The breath is directed forward over the snout instead of straight up into the air, as with larger whales. This means it is virtually impossible

When 160 pilot whales beached at Ft. George Inlet, Florida, officials called in the Smithsonian's James Mead. Armed with computer profiles, Dr. Mead and locally drafted volunteers rushed down to the beach to take critical measurements before decomposition took its toll. Random autopsies were also performed in an attempt to isolate the factors that cause such strandings, which are surprisingly common and still poorly understood.

The Magnificent Foragers

to observe them at sea unless you get extremely close to one."

Mead's data confirms that the Blainville's beaked whale is only one of many kinds of whales that live along the U.S. Atlantic coast in greater numbers than anyone suspected. Also, many species once believed restricted to the Pacific may dwell in the Atlantic as well. It is becoming clear that some of these animals are very regular in their movements. Mead is assembling information on the size and distribution of their populations, and on their eating and breeding habits.

"We do know for a fact that at least 36 different kinds of whales roam the northern Atlantic. Some travel along the coast in large schools, but others prefer solitude, sticking to deep-ocean areas and diving to depths of 50 to 199 fathoms. They stay below for up to an hour and then surface for only a moment."

The lives of such whales are for the most part a closed book. To help open it, Smithsonian scientists have made beach pickups of marine mammal remains for more than a century. Until recently, the pickups were random windfalls. When scientists at the Museum heard about a stranding near Washington, D.C., they would drive their cars to the beach and do their best to cram some study material into the car trunks. Their cars would be the objects of stares and gasps as they wobbled down the road, the flukes of a 9-foot, 500-pound whale sticking out at the rear like a giant rudder.

With a rented crane, a Smithsonian team hoists a humpback whale, one of the baleen whales, onto a flat-bed truck. Specimens are frequently brought back to the Institution and kept in deep-freeze lockers since full autopsies are difficult to perform in the field. In addition to examining such elements as parasite load and stomach contents, scientists test for heavy metal concentration. At the top of the marine food chain, whales concentrate these pollution-indicating heavy metals in tissues and organs.

The Solitary Researcher

A pilot whale fetus is examined at Ft. George Inlet. Relative to other mammals, the whale is still a mysterious creature, living its life at depths inaccessible to man. Studying fetal development illuminates species evolution patterns.

Now specimens no longer dribble in by chance. Mead has recruited dozens of volunteer whale watchers, Coast Guardsmen, fishermen, and park rangers to help. One dedicated naturalist, a high school biology teacher on Long Island, encouraged his students to watch for strandings along the coast. When volunteers learn of a marine mammal aground, they call the Smithsonian's Scientific Event Alert Network (SEAN), which immediately passes the word to Mead.

"Strandings happen most frequently during the winter when the Atlantic gets rough," he says. "If they occur between Cape Cod and Cape Hatteras—within a day's drive of the Smithsonian—my assistant and I go to look at them in a special flat-bed truck, equipped with a winch and a hydraulic system. For more distant strandings I call volunteers who live close to the scene. They'll examine and photograph the dead animal, then air-freight frozen samples to the Smithsonian. If it turns out to be rare and unusual material, I'll fly to the scene. We recently traveled to Panama and Costa Rica to investigate a beaked whale stranding and a mass stranding of dolphins that occurred at about the same time."

Mead and his colleagues now retrieve more than 100 marine mammals a year, a high percentage of the total strandings along the Atlantic seaboard. Getting on the beach in time to do an autopsy enables them to accurately determine the cause of death and take samples for analysis of heavy metal residues, pesticides, and other pollutants. They also check the animal's stomach contents and reproductive condition.

"I've been very fortunate so far," confesses Mead. "I tend to be lucky at finding whales. A year ago I went out to San Diego for a conference and a beaked whale beached there the next day. It was the first time since 1946 that a beaked whale had stranded in that vicinity.

"Not long after, I visited Argentina and happened to walk out to the beach. There was this very strange-looking stranded beaked whale. It turned out to be a freshly beached *Tasmacetus shepherdi,* probably the least known of all marine mammals, a kind previously seen only in New Zealand. Sometimes it's almost uncanny. I long ago gave up taking a vacation at the beach for rest and relaxation, because every time I did a dead whale or dolphin washed ashore."

Right, mammalogist Mead examines a rarely seen beaked whale, Tasmacetus shepherdi. *It is one of the whales about which so little is known that a single beaching provides a mass of critical new data. Opposite, the strands of which sea myths are made: long dead, this bottle-nosed dolphin is almost unrecognizable and could easily result in a layman's report of a "monster" stranding.*

The Magnificent Foragers

Sea Birds

The Antarctic

The Smithsonian's George Watson studies the distribution, systemics, and ecology of marine birds, particularly those that live in the Arctic and Antarctic regions. Most of Dr. Watson's recent work has concentrated on the Antarctic, and he has published the first comprehensive field identification guide to penguins, petrels, and other resident and vagrant birds of this area. To complete this project, he made three collecting cruises in sub-Antarctic and Antarctic waters, researched pertinent literature—some of which dated back to Captain Cook's time—and examined specimens in museum collections, including those at the Smithsonian gathered both by the U.S. Exploring Expedition of 1838-42 and Admiral Richard Byrd's U.S. Antarctic Service Expeditions of the 1930s and 1940s. His book standardizes nomenclature for Antarctic sea birds and benefits tourists as well as scientists.

A South Polar skua chick, a predator gull, on Litchfield Island, above left. Above right, an immature glaucous gull snatches a scrap of food from the sea's surface. Right, George Watson holds shut the formidable bill of a giant fulmar while banding it. He and other researchers banded these birds on Cormorant Island to trace their movements and to establish a marked population for future studies. Opposite, an adult glaucous gull circles the U.S. Coast Guard icebreaker Glacier, looking for an arctic cod that might appear when the bow wash overturns a piece of ice.

III.   Two Islands

One is a tropical rain forest lying in the middle of the Panama Canal, the other a mere speck of coral reef in the Caribbean. But they are "scientific" islands in the sense that each has been scrutinized—painstakingly and even exquisitely—by a broad spectrum of Smithsonian scientists.

Barro Colorado

Gatun Lake, Panama

Barro Colorado is an island in the middle of a lake through which the Panama Canal runs. Early in this century, the engineers who built that convenient link between the Atlantic and the Pacific dammed the Chagres River to create Gatun Lake, about midway through the 50-mile Isthmus. When they did, river waters flooded the Chagres valley, though several of its highest hills protruded above the surface of the new man-made lake. Of these instant islands, the largest was Barro Colorado. Named for its red soil, the 6,000-acre island was set aside as a reserve in 1923 and used exclusively for biological study. Since then, distinguished naturalists from all over the world have lived on the island and studied its profusion of tropical plants and animals. In 1946 this research came under the auspices of the Smithsonian Tropical Research Institute, a bureau of the Smithsonian that operates several other field research stations in the Canal Zone.

One reason Barro Colorado is a very special place scientifically is its location midway between two major continents—with elements of both North and South American plant life. Also, because it has been under study so long, scientists can draw upon and synthesize an immense accumulation of data on how the island's life has intermingled, adjusted, and evolved.

Arriving at Barro Colorado's boat dock, scientists and visitors must first climb a flight of 200 stairs to a ridgetop. There, in a clearing between two streams, sits a tight cluster of buildings—laboratories, an animal house, dormitories, and a dining room with attached kitchen. The station is surrounded by tropical forest.

Although visitors can work in comfort in Barro Colorado's air-conditioned labs and listen to the music of Vivaldi, it is difficult to filter out the presence of animals and the forest. The cries of howler

Visiting nuns on their way to the Smithsonian's tropical research station at Barro Colorado. Besides hosting scientists from a range of disciplines, the station is open to visitors who may stroll along a sign-marked trail into the forest. An aerial view at right shows the research compound in the clearing at the top of the hill.

monkeys, hawks, parrots, and other birds reverberate through the trees. Spider monkeys swing on the electrical wires on their way through the clearing; tapirs break into the kitchen to steal bread; at night, a bewildering variety of moths fly around the lights and settle on the screens. The forest, too, continually intrudes on the little clearing. Trees fall, crushing cages and threatening buildings; new plant growth must be constantly cut away with machetes, indispensable tool of the tropics.

When scientists go into the forest, they notice that it is cooler and more humid than in the clearing, particularly in the middle of the day. They cannot see far in any direction. Frustratingly, they can hear and smell more than they can see. Bird songs pierce down from the upper canopies, the fragrance of blossoming flowers wafts down from somewhere high above the forest floor. Very well, then, take science to the canopy. An observation tower now looms up above the top-most layer, enabling researchers to watch and eavesdrop on birds and insects and to examine petals and fruits. Turning their eyes toward the lake from this vantage point, they can see the ships of the world passing through the canal—silent, commercial, oddly monstrous and gawdy against the green, tropical backdrop.

Returning to the clearing after a day in the tower, scientists contribute their observations to dinner conversation: what trees are in flower, what fruits are ripening, or what animals they spotted. Others, who have been studying the fish in the lake or sifting through dead leaves on the forest floor searching for insects, have different tales to tell. Still others will not be at dinner at all, for they are out in the bush trying to study animals active at night. The stories of the "bat people" and other nocturnal rovers will enliven breakfast.

The Magnificent Foragers

Like the sea, the tropical forest is layered with characteristic life forms at different altitudes, from the forest floor to the top-most tree canopy. Shown here, the view down from an observation platform erected high in a Tachigalia *tree.*

Many of the recent visitors at Barro Colorado are participants in an environmental program that is drawing on the talents of scientists from other Smithsonian bureaus as well as those from universities and government agencies. The program is centered near the principal station at a 25-acre research site where instruments constantly measure and record rainfall, sunlight, temperature, and humidity. Additional equipment calculates plant growth rates, water flow, and soil composition. Insect traps are scattered about to determine seasonal variations in insect numbers and species—one of many biological monitors connected with the overall scientific effort.

Barro Colorado is inhabited by innumerable insects, about 300 species of birds, nearly 2,000 different flowering plants, and close to 100 species of mammals, including 40 kinds of bats. Among the other mammals are ocelots, peccaries, three-toed sloths, tapirs, anteaters, and a variety of spectacularly active monkeys. Except for mosquitoes and ticks, none of the island's animals may be harmed; botanists are not even allowed to collect plants. As part of approved projects, however, animals may be trapped and tagged. But then, after a minimal confinement, they must be returned to the forest and released.

Richard Thorington, a Museum mammalogist and an authority on New World primates, makes regular trips to Barro Colorado to study the island's howler monkeys and their food resources. Howlers, along with cebus monkeys, red-naped tamarins, night monkeys, and spider monkeys, make up the island's resident primates. Because these animals are not hunted, they are relatively unafraid of humans and thus are much easier to approach than is usually the case in the wild.

Much of the existing data on the natural history of these animals was gathered at Barro Colorado. Among the notable research in this field was Raymond Carpenter's pioneering Barro Colorado studies of howler monkey behavior in the 1930s and the work of the Smithsonian's Martin Moynihan, who studied the island's red-naped tamarins, examining in particular their methods of communication and their relationships with other animals.

Early in the morning when he is on Barro Colorado, Dr. Thorington hikes into the forest in search of howler monkeys. Howlers, the largest monkeys on the island, weigh as much as 22 pounds. Troops move slowly through the treetops, climbing and walking on all fours, clinging to branches by their long, prehensile tails, and pausing to feed on young leaves, buds, flowers, and fruits—especially figs and hog plums. "I can usually locate them easily," says Thorington. "During the day, male howlers advertise a troop's presence by deep roars that can be heard for miles."

Falling branches, low-flying airplanes, rainstorms, and trespassing humans all bring howls of protest. Most dramatic of all is the chorus

Researcher Sally Levings faces off eye-to-eye with a resident turtle. Below, scientists, students, and visitors at dinner. With an international and diverse group typically present, meals take place in a congenial, stimulating atmosphere, one of the special aspects of research at Barro Colorado.

Two Islands

Spider monkeys were introduced to the island in 1959 in an effort to re-establish indigenous but absent wildlife; over-hunted, spider monkeys had vanished from the region. The program has not been an unqualified success, however, since breeding females have produced a preponderance of male offspring.

Bloat-bellied Bufo marinus sits straining at his banding girdle, striped with unique colors by George Zug to identify it. The largest of the marine toads, its name is deceiving; it is not a salt-water toad, but a denizen of tropical savannah cane fields. Feeding primarily on cane beetles, marine toads were exported to Hawaii from Puerto Rico in 1934 for biological pest control in the sugar cane fields. They successfully adapted and spread throughout the South Pacific.

that pours forth when two or more howler troops that covet the same fruit approach and come face to face. The males break into loud, deep roars, sung in concert with high-pitched calls of the females and juveniles. Usually one of the troops moves in on the desired fruit, either displacing another troop or blocking its access to the food. Although there is seldom physical combat, many males bear large scars, suggesting to Thorington that these intertroop encounters are not all just bluff and roar.

Thorington uses an anesthesia dart gun to bring the monkeys down from the trees. The drug causes them to wobble, fall, and drop into a net. Thorington then takes body measurements, blood samples, toothcasts, and fingerprints. The tails are marked so that he can identify individuals later. The howlers soon revive and scramble back up into the trees.

The fruit trees on which the monkeys feed are also being marked and mapped in the course of the study. The project is expected to provide the first complete picture of a howler troop's long-term population dynamics and the food resources that influence them. With a food supply that is apparently stable, Barro Colorado's monkey population has been growing by leaps and bounds, from a few hundred in the early 1950s to more than 1,500 by 1975. Nevertheless, threats to this large population do exist. A yellow fever epidemic, for example, swept through Central America in 1950, killing a large percentage of the Barro Colorado population. Thorington fears the disease may well strike again.

"There are also the gradual and inexorable changes that are occurring in the monkeys' forest domain," he adds. "Botanists who have examined the island's fig and hog plum trees have noticed that the trees do not seem to be producing seedlings, which leads me to believe that the howlers' food supply will likely dwindle in the years ahead."

Other mammals on the island being caught and marked by scientists to amass data on their life histories are the coati, a long-nosed animal related to the raccoon; pacas and agoutis, large South American rodents; and the red-tailed squirrel. The squirrel study is being conducted by Thorington and a student, Larry Heaney, both because it relates to the monkey work and because squirrels are one of their special interests.

"On my first visit to Barro Colorado in the mid-1960s, squirrels were rare and I saw only two individuals," Thorington explains. "By the early 1970s, however, the squirrel population had increased dramatically. I was able to see more than a half dozen squirrels an hour. It occurred to me that they might be feeding on some of the same foods as the howlers. Therefore I wanted to try to find out something about their natural history."

Thorington was already well acquainted with the red-tailed squirrel's North American relative, the eastern gray squirrel, which he had studied for a number of years in his Bethesda, Maryland, back yard. There, Thorington had developed techniques for marking the animals so he could keep track of them.

The same techniques came in handy on Barro Colorado. Captured alive in traps, each squirrel was given an ear tag and a collar of plastic beads, individually color-coded so that each squirrel had its own combination. When released, the marked squirrel could be

Howler monkeys studied by Richard Thorington are the most common primate on Barro Colorado and the dominant arboreal herbivores. They move through the canopy feeding on figs and plums, spreading fig seeds over the island. The seeds are not germinating, however; as the island matures, forest cover becomes denser and the fig seeds cannot germinate in the deep shade.

Preparing for night collecting, entomologist Nicholas Smythe adjusts ultraviolet insect traps. Below right, a newly tagged red squirrel recovers from the mild tranquilizer administered during capture. Mammalogist Richard Thorington studies the ecology of squirrels to determine if they compete with howlers for food.

watched in the forest, with particular attention paid to its social activities and feeding habits.

Thorington found that the red-tailed squirrels behaved in much the same way as the gray squirrels back in Maryland. Both feed predominantly on hard seeds but differ in that the gray squirrel eats hickory nuts and the red-tailed eats palm nuts. Both squirrels scatter-hoard these seeds in holes in the ground, though the red-tailed squirrels also temporarily store them in the crotches of trees.

As for their impact on the forest, Thorington soon learned that the howlers and the squirrels consumed different foods. But, interestingly, they did so in such a way as to have an exactly opposite effect on the forest—one positive, the other negative. Since the howlers eliminate the seeds of the tree fruits they eat, the seeds may subsequently germinate and grow. The red-tailed squirrels, on the other hand, chew up the seeds, destroying them. It is a potential problem for the forest when the squirrel population becomes as dense as it is now, according to Thorington. When there are several squirrels per hectare of forest, they destroy a high percentage of the seed crops of some of the palms and leguminous trees in which they feed.

Thorington does not yet know why the Barro Colorado squirrel population is exploding. He does point out, however, that squirrels are tremendously fecund when conditions are favorable, that they are susceptible to mange and many other parasites and diseases, and that sharp fluctuations in their populations have been documented in other parts of the world. A good chance exists, then, that the squirrel population of Barro Colorado faces a decline.

* * *

With its high humidity and heat, Barro Colorado provides ideal conditions for dozens of different amphibians and reptiles. Often out at night crawling, swimming, or hopping, all are candidates for scientific study. Some were not present on Barro Colorado Island when it was originally created. More of the island initially bore open fields and secondary forest. As these areas have returned to forest, a few species have become extinct while several mainland species have arrived, either by swimming to the island or rafting there on floating debris.

Enormous numbers of frogs breed in the many ponds and streams,

Beauty is in the eyes of the beholder. To the Museum's Charles O. Handley, this Mimon crenulatum, *with its rose-flushed ears and nose, is beautiful. One of the "whispering bats," its scalloped-edged nose "leaf" is thought to guide sound waves back to the ears. Dr. Handley is conducting the first long-term population study of marked bats.*

and in the evening the calls of the male frogs join with other animal cries, producing a fantastic chorus. The Smithsonian's Stanley Rand has studied behavioral interactions of chorusing frogs. When a male calls to attract females for mating, he is sending a message packed with information. He gives the female his location, tells her what species he is, and makes it clear that he is ready to mate—perhaps boasting of his qualifications to do so. Moreover, he is able to make all of this heard over the general uproar of other voices all around him.

Young frogs living in the island's system of stream-bed ponds share their environment in such a way that each species can survive and get the most out of its habitat. Museum reptile and amphibian specialist Ronald Heyer gathers data on the "partitioning" of these ponds that helps him understand how individual tadpole species interact within a community setting. Dr. Heyer visited Barro Colorado one year during the dry season—February and March—and again during the wet season in July of the same year. Tramping along the stream beds, dip net in hand, he searched for ponds, many of which were no larger than a soup bowl.

In some of the ponds where he found no tadpoles, Heyer instead found fish—tadpole predators. Establishing themselves in fishless ponds was obviously crucial to tadpole survival. Heyer found one species that solved this problem by laying eggs only in ponds smaller than those usually occupied by fish. Another successful strategy involved laying eggs in a sufficient number of ponds to exploit the patchy occurrence of fish in the pond system.

Heyer discovered that Barro Colorado Island frogs had seasonal breeding periods resulting in a minimum of conflict for the limited food resources. Most of the ponds contained only a single species of tadpole. Of the five different tadpole species Heyer located, two occurred in ponds exclusively during the dry season, and two exclusively in the wet season. The fifth tadpole species lived with the others, but it did not compete with them directly for food because its large funnel

Two Islands

A metallic-colored arboreal ground beetle of the genus Calleida. Entomologist Terry Erwin is studying this apparent contradiction in terms. As it turns out, the beetle is not confused, the forest is. The tropical forest canopy supports a terrestrial environment high above the forest floor. In the humid heat, organic matter in the crooks of trees decomposes rapidly, creating pockets of soil which support plant life, reptiles, and ground beetles. In fact, almost half of the known species of ground beetles in Panama are arboreal.

mouth enabled it to consume food from the pond's surface film. In contrast, other tadpoles possessed mouthparts adapted to scraping and chewing so that they could feed on the bottom of the pond.

The giant toad, an extremely adaptable amphibian that is at home in tropical clearings throughout most of the world, was recently studied on Barro Colorado by the Museum's George Zug.

During the first two or three hours of evening, Dr. Zug was able to pick up the toads easily outside the island's cottages and laboratory buildings, where they prey on beetles and other insects attracted to light. Before releasing the toads, Zug would take notes on their location, sex, and size. Then he would fasten a colored identification belt around their stomachs, making it possible to keep track of the frequency of their feeding. He discovered that their dining tended to be irregular. Often after a giant toad had eaten its fill of beetles—its bulging stomach crackling when Zug palpitated it—the toad would hop to its burrow and sleep for several nights before coming back for more food.

"Particularly surprising is the apparent short life span of most giant toads," says Zug. "They live hardly more than a year. The adults have poison glands that keep predators away, so it is unlikely that they are being eaten. What is happening, I believe, is that the toads are becoming fatally dehydrated during the island's dry season."

Charles O. Handley's bat survey is another example of night work on Barro Colorado. In this pioneering effort, Dr. Handley is marking the different generations of bats roosting on the island and following up on these flying mammals through their long lives of up to 35 years. He charts their movements, reproductive history, seasonal variation in population and distribution, social structure, and feeding habits.

Tropical bats have strikingly different ways of feeding. Some are vegetarian, chewing up fruits and imbibing the juices. Others catch insects, while still others prey on small animals. Among the latter are bats that forage the forest floor for rats or lizards, and bats that capture minnows and other small stream fish. There are also a few species of vampire bats that bite forest animals or birds and drink their blood.

*　*　*

Walking along the system of trails on Barro Colorado, visitors are constantly reminded of insects, the dominant form of life in the tropics. Again, the reminders are mostly auditory: the ever-present din of cicadas and katydids, the occasional buzzing of an orchid bee or metallic woodboring beetle, or the faint crackling of leaves beneath the feet of millions of army ants on the move, hunting for food. But

Sally Levings staking out perimeters for a study of fauna of the forest floor.

on the island there are also reminders of another dominating life form—biologists. Many of the trails and streams are named after prominent early entomologists, among them William Morton Wheeler, T. C. Schnierla, R. C. Shannon, F. E. Lutz, and James Zetek. Dr. Wheeler, an ant specialist at Harvard, spent long periods on the island, as did Dr. Schnierla, an animal behaviorist long associated with the American Museum of Natural History and producer of a famous study of army ants.

The impact of army ants and their relatives on ground beetles (*Carabidae*) is one of the prime interests of Terry Erwin, a Museum entomologist. Through his studies at Barro Colorado and other forest sites in both the Old and New World, several hundred tropical and temperate ground beetles have had their natural history observed and recorded, most for the first time.

It was on Barro Colorado that Dr. Erwin and his wife LaVerne first set out to study tropical ground beetles. "They are not easy to find," he admits, "because they live in a patchy distribution on the forest floor, unlike the situation in the north where they are distributed more evenly. Here you must search for a patch to begin with. We discovered that during the dry season on the island, ground beetles bury themselves in the deep piles of leaves under the crowns of recently fallen trees. In May, as the rainy season begins, the beetles forego the leaves and move to places on the forest floor where fruit or blossoms have fallen in such profusion that fermentation is under way. Some beetles eat fly larvae (maggots), some eat the fruit seeds, and some eat other beetles. After eating, mating, and egg laying, one generation dies and the new one flies back to a fallen tree to wait out the oncoming dry season.

"However," he goes on, "of the 270 species of ground beetles inhabiting Barro Colorado today, only one-fourth of them live on the forest floor. More than half live in the forest canopy, one hundred feet overhead. We are just now beginning to study these arboreal species."

Photography is an important scientific tool at Barro Colorado for Neal Smith, who studies bird populations and their migratory habits. Part of his technique involves censusing migrants through photographs. At right, migrating hawks.

The Magnificent Foragers

Some 300 species of birds have been observed on Barro Colorado and its adjoining waters. Alexander Wetmore's *The Birds of the Republic of Panama,* a monumental study of what is known about native Panamanian birds and other migrators using the narrow flyway of the Isthmus of Panama, is based partially on observations made at Barro Colorado. One of the ornithologists who contributed to the observations was Frank Chapman, who in 1929 made the first published reference to a spectacular migration of turkey vultures over Gatun Lake. "They came from as far as one could see toward the north," he noted, "in loose flocks of 25 to 50, separated by short intervals, and took about half an hour to pass."

Vultures pour into Central America from the North in tremendous flocks in October. Wetmore noted that while many remain on the Isthmus, thousands pass farther south into northwestern South America, returning in February, March, and early April. Along with the vultures come thousands of Swainson's and broad-winged hawks, following a definite pattern of flight. The resulting concentration of birds of prey is unequalled anywhere else in the world. Smithsonian scientist Neal Smith monitors the population fluctuations of these hordes as they pass over the Canal Zone. He was able to count the actual numbers of each species by photographing the stream of birds in a non-overlapping manner and employing an electronic scanning counter. Dr. Smith's studies were undertaken with the cooperation of the U.S. Air Force, which considers flocks of hawks and vultures to be a navigational hazard and wants to be able to predict when, where, and at what altitude the birds are likely to be flying. Amazingly, Smith found that, faced with rainstorms, the hawks and vultures climb up over the storm rather than land. Authenticated sightings of hawk and vulture flights 5,760 meters over the Canal Zone have taken place. At these altitudes the birds do not flap their wings, but ride the convection currents atop the storm, propelling themselves at speeds exceeding 40 knots.

Barro Colorado's resident black and turkey vultures were studied by Laurie A. McHargue, who kept a photographic record of their nesting habits and observed development of both young vultures and adults. Other major ornithological studies on the island include Edwin Willis' study of five or six species of ant-following birds that travel with raiding columns of army ants, feeding not on the ants, but on the insects that are flushed by the ants' advancing columns. Smithsonian ornithologist Paul Slud works in the New World tropics correlating ecological background with bird census observations. His analysis of this information establishes links between the comparative distribution of birds and prevailing environmental and climatic conditions.

On a field trip to Barro Colorado to survey its bird population, Dr. Slud walked the island's trails with little more than field glasses and a notebook. Because of his familiarity with the birds of the Central American area, he did not need to trap the birds in mist nets to make identifications. He did it by looking and listening. Whether he located the bird in the tree-top foliage or close to the ground depended on the species. Some birds can be found almost anywhere, but a great many confine themselves to just one level of the forest. Most of the vividly colored birds such as the toucans and cotingas stay high in the trees. The majority of the birds that live at the lower level, however,

Pendulous nests of the oropendola, a tropical bird, hang from a Barro Colorado tree. Each nest is the site of a complex, four-way relationship involving oropendolas, giant cowbirds (brood parasites that lay their eggs in the nests of other birds), botflies (also parasites), and stingless bees (which act as a parasite deterrent). The arrangement is unusual in that the presence of the brood parasite cowbird sometimes benefits the host oropendola. Opposite, grounded by a thunderstorm, vultures wait for large thermals to build up again to carry them up and over the storm.

tend to be drab, blending in with the brown of the forest floor.

Slud's work on the island focused on identifying permanent residents so he could compare their population with that of the mainland. Changing ecological conditions, he found, were influencing the island's bird population. For instance, the island is becoming more heavily forested and closed in. Birds that prefer openings are becoming less common.

"One thing that struck me," says Slud about his visit, "was that although Barro Colorado is an island on an inland lake, not more than a mile or two from the mainland at some points, its bird population showed many of the same effects of isolation that you see on remote oceanic islands. There were imbalances in the populations of the birds living on the island. Compared to mainland situations, some species were very rare, others were superabundant, and a number of others were absent altogether."

For Lloyd Ingles, another scientist who worked on the island, the music of Barro Colorado's many birds and other wild animals provided a soothing and pleasing ending to a day of research. Many years ago he wrote: "Each evening about sundown the big chestnut-mandibled toucans mount the top branches of the tallest trees and with much bowing and waving of their enormous beaks, break the solitude with their loud squawks and yippings. One does not mind this, however, because he has learned that the next number on the program will be the long, wailing, flute-like notes of the great tinamou which serve as vespers every evening on Barro Colorado. When the tinamou has sung his song, the nocturnal chorus of the insects and amphibians has already begun. Many bats are darting about, and the mellow hoots of a spectacled owl greet the night from the dark forest. There should be more Barro Colorados."

Carrie Bow Cay

Off Belize, Central America

Rich in underwater plant growth and honeycombed with innumerable refuges for marine animals, coral reefs are the most productive of the world's ecological communities. Their natural architecture nurtures fish, lobsters, crabs, shrimp, and clouds of smaller creatures. Yet an increasing number of reefs the world over are in trouble. When natural forces become unbalanced, the reef-building organisms die. Worms and other burrowing creatures weaken graceful walls of coral stone, and the force of waves completes the reef's destruction.

What causes this deterioration? Is man's pollution at fault? Or are the changes caused by ecological fluctuations that have nothing whatsoever to do with man's presence? As yet, the answers are unclear because so little is known about reefs. All the plants and animals that live on them are still not inventoried. The different ways in which they relate to each other have hardly been explored. Nor is the impact of such variables as light, temperature, and wave action on reef organisms as yet thoroughly understood. In addition, facts are missing on reef population changes as well as on how reefs budget energy, build up, and break down.

To obtain the complex data required for an answer, the Smithsonian decided to look for the ideal coral reef where a group of its scientists could undertake a long-term study.

"Reef-building coral animals and plants do not grow prolifically in waters where temperatures fall below 80 degrees F.," notes zoologist Klaus Ruetzler. "As a result, reefs are seldom found off coasts where cold currents sweep by. The subtropical and tropical Caribbean and the Indo-Pacific are the great coral seas. We decided on the Caribbean because it is closer to home and because the data collected there would aid in the conservation of Florida's endangered reefs."

Sites on Acklins Island in the Bahamas, St. Croix, the Virgin Islands, and Discovery Bay, Jamaica, were examined in detail and, while each had virtues, they were all eliminated. It soon became evident that the team's requirements could best be met on the magnificent barrier reef off Belize, in Central America. It runs for almost the entire length of Belize's 170-mile-long coastline. Only Australia's Great Barrier Reef is longer.

The barrier reef strip selected is made up of a series of separate coral reefs, some of which are very small. Others are many miles long and quite wide, separated by deep channels. On the land side lies a lagoon, 10 to 15 miles wide at points, and on the seaward margin, the deep Caribbean.

Hundreds of cays, or keys, rise behind the Belize reefs. These small islands were born of silt and coral sand, washed and blown onto a reef site forming shoals on which mangrove trees and other vegetation could take root. Some of these islands became lushly vegetated, with great palms towering above white beaches. Others are simply sandbars, exposed at high tides, or clumps of mangrove trees with no dry land.

Plants colonize these cays rather easily. The seeds wash ashore on floating debris, or birds drop them. One of the Smithsonian's botanists, Raymond Fosberg, studied the cays of Belize and found that the larger

Opposite, looking out at the reef crest from the windward side of Carrie Bow Cay, site of one of the Smithsonian's marine laboratories.

The Magnificent Foragers

islands supported as many as 35 to 40 different species of land plants. "The vegetation stabilizes the cays and keeps them from being washed away by the buffeting of violent storms," Dr. Ruetzler points out. "Still, each time a hurricane strikes the area, many of the islands become a little smaller. Some eventually melt away and submerge."

The Museum team judged Carrie Bow Cay, one of the small islands, as best suited for a study site. It has only a scattering of palms and sparse vegetation, and is hardly more than 50 paces wide at some points. But its reef was just what the scientists wanted. Vigorously developed and unpolluted, it slopes off steeply into the ocean, giving it a compactness that enables researchers to move easily from one biological life zone to another.

A visiting scientist dries seaweed, top, prior to chemical analysis in the cay's laboratory, opposite. Carrie Bow facilities are often open to scientists from other institutions. The dugout canoes, below, used by Belizean fishermen, lie on the leeward shore of the cay.

One of the first tasks at Carrie Bow Cay involved laying down a permanent transect cable as a marker for experimental work. It extends from the lagoon on the landward side of the cay, through 900 feet of shallow water covering the flat part of the reef out to the crest, where the sea's surf breaks. From the crest, the line gently follows the reef floor downhill for about a quarter of a mile, reaching the reef's outer rim and a steep dropoff of nearly 1,000 feet.

The transect line passes through all of the reef's four biological zones, characteristic forms of life crowding each one. In the lagoon, bivalves and other mollusks burrow into the muddy bottom amidst the turtle grass. In shallow back-reef waters, crabs, sea urchins, and other echinoderms hide under coral patches on the sand or on other strong rubble. On the reef crest itself, soft growths—both plants and ani-

The Magnificent Foragers

Klaus Ruetzler installs asbestos gutters on a reef slope at a depth of 50 feet. The gutters serve as artificial caves where marine invertebrates can settle. Dr. Ruetzler's experiment will provide data on competition among the reef animals for this living space: which animal will take over the area first?

mals—cling to the coral rock where the waves crash in and flow over into the lagoon.

Finally, waters deepen on the seaward side of the reef crest. Here schools of red, yellow, blue, and green finger-size fish flit through the blue water, browsing among forests of branched staghorn coral. Clusters of magenta fans—gorgonians—sway back and forth in the currents. Sharks glide by. Eagle rays pass, slowly undulating their wing-like fins. Brown sponges resembling enormous Greek urns grow on the bottom, and all around them in the chambers of the coral bed lie thousands of tiny animals: snails, nudibranches, tubeworms, and brittle stars—all seeking food and shelter in the densely populated and complexly intermeshed reef world.

Wherever a fragment of coral is killed by predation or an occasional storm, numerous small plants (algae) are quick to settle. Some contain calcium and aid in the reef-building process. Most are small and uncalcified, in some places forming thick turf over everything not otherwise occupied by an animal. These algae provide much of the food for the reef. Parrotfish, sturgeon, damselfish, and innumerable invertebrate animals graze on this submerged meadow.

"After surveying the reef all the way to where the corals and algae thin out in the deep waters off the reef rim," explains Ruetzler, "we began to take regular measurements of the reef water's submarine light intensity, salinity, acid-alkaline balance, oxygen, turbidity, and motion. When correlated with long-range meteorological data, these factors will help us determine the influence of tides, cloud cover, wind force and direction, wave action, and other environmental conditions on the growth rate of a coral community."

The Magnificent Foragers

Ruetzler explores the weird under-water forest of reef gorgonians and sponges off Carrie Bow Cay.

Two Islands

The Smithsonian's Ian Macintyre conducted experiments at Carrie Bow Cay to demonstrate how one of these factors, light intensity, can be critical to coral growth.

"At regular intervals," says Dr. Macintyre, "a colleague and I placed plastic bags over coral heads and released a red dye inside, staining the exposed coral. Eventually the corals were collected, sectioned, and x-rayed. The stain patterns seen on the x-rays enabled the corals' growth history to be plotted. Then a relationship could be established between the growth, light intensity, and distribution, as measured in the corals' underwater habitat. Light frequently proved to be the principal factor in coral growth and form. To learn more about environmental control of coral growth," he adds, "we are now transplanting reef corals to sites with quite different light conditions."

Much of Ruetzler's work at Carrie Bow Cay has concentrated on sponges. These marine animals have developed a symbiotic arrangement with microscopic algae in the water, and Ruetzler investigates the beneficial aspects of that relationship.

"The sponges draw in water and filter out single-celled blue-green algae, which then grow within the sponge's cellular system," he explains. "Algae thrive here because they can manufacture starch from the waste carbon dioxide produced by the sponge. It's almost as if the sponge had a vegetable garden. It can harvest nutritious substances the algae give off. I feed the sponges algae material that has been labeled with radioactive chemicals. Later, I take small tissue slices. That way I can follow the digestion of the particles on an ultrastructural level, visible with the electron and scanning electron microscopes at the Museum."

One of the many little-known organisms that live on a reef is the sea fan. When sea fans are collected, they are turned over to Frederick Bayer, a leading authority on these delicate shrub-like reef inhabitants. As many as 100 different kinds of gorgonian sea fans and related organisms live on an average Caribbean reef. A few are distinctive, but many defy precise identification. To complicate matters, a single sea fan species doesn't look the same in different locations on the reef where it reacts to different wave motion or light.

"For instance," Dr. Bayer explains, "one kind might sprawl all over the place in one situation, and another become a compact shrub in another. I develop data that will make it possible to identify species, regardless of their ecological variation. The task is far from complete, but once it is accomplished we can go on to investigate other such practical problems as the effects of pollutants and sediments on their physiology."

In dense concentrations, the gorgonian plant-like form provides cover for hosts of fishes and other active organisms. If the gorgonians weren't there to hide among, many of the fishes wouldn't be there either. In other words, anything that damages the habitat damages the inhabitants. Gorgonians are extremely sensitive, so increased sedimentation from dredging, freshwater run-off from cleared land, or pollution from human habitation can destroy them.

Porter Kier, Museum Director and a paleontologist, does his own diving to study Carrie Bow Cay's echinoderms—especially sea urchins and sand dollars—to find out what kinds live on the reef, what they eat and what eats them, their energy budget, how they generally relate to coral reefs, and how their distribution changes during the year.

Belizean fishermen bring in colorful catch taken off the barrier reef.

The Magnificent Foragers

The spiny sea urchin uses its spines for locomotion, for lodging itself in crevices on the reef, and for keeping predators away.

"When I first began my study," he reports, "there were a lot of sand dollars on the reef flats. Now there are none. We know that it isn't because of human disturbance, which is at a minimum at Carrie Bow Cay. So what happened? Where did they go? It's possible that it's the result of a storm. On the other hand, it may be that the population is just growing old and dying."

Some of the sea urchins collected by Dr. Kier live buried 8 inches down in the mud. They construct a tube through which they get their oxygen. "It's useful to learn how the same structure that I see in a fossil is used by a living animal," says Kier. "From observations like this you can see what advantage there is to a certain kind of modern feature, and why it is that the animal evolved as it did."

The earliest sea urchins lived above the sand on the ocean floor. They used their tube feet—the tiny tentacles that poke from their shells—to walk and to obtain food and oxygen. But none of their tube feet were specialized. Then, about 250 million years ago, great environmental changes took place, and a new group of urchins evolved.

"In these more advanced animals," notes Kier, "one set of tube feet does the food gathering, another does the job of walking and holding onto rocks. A third set takes care of respiration. Because they have special oxygen-getting tube feet, like gills, the flat sand dollars were able to burrow and live in the sand. Here they could hide better from fish and other predators, which is why they survived over the course of evolution."

Among the destructive forces attacking the reef are some burrowing animals known as sipunculan worms. The activities of these marine animals fall into the province of the Smithsonian's Mary Rice. Incongruously enough, in the tropical island setting of Carrie Bow Cay, she uses a geology pick and a chisel. "Someone dives to collect coral rocks, identifies the material, and brings it to me," recounts Dr. Rice, an invertebrate zoologist. "I then sit all day and slowly break it up, extracting the worms, identifying them, and weighing them to determine how densely they inhabit the rock. I'm trying to relate density of sipunculans to different types of coral rock, to see if there are some kinds of rock in which they bore more readily than others.

"I'm also interested in the mechanism such worms use to bore into the rocks," she continues. "This long and slow process is poorly understood. They enter into tiny crevices when they are small, extending their anterior outside to feed on algae and other debris that adheres to the rock. As they grow, they enlarge the crevice, softening it by chemical means and rubbing against it with their rough cuticle. Once they're in the rock they're in it for a lifetime. They never leave. We've been looking at their burrows at the Smithsonian under a scanning electron microscope, but you can't study the worms in action in the lab. Once we take them away from the reef they won't bore anymore."

Thomas Waller, a paleontologist, also has made many dives at Carrie Bow Cay to study its scallops, clams, and other mollusks. According to Dr. Waller, one of the beauties of working on the cay is that in one dive a researcher can go from the deep water along the vertical reef rim up to the reef crest and, moving from one zone into another, see the changes that take place over the range of mollusk life.

"Because our little lab on the cay is so close to where we collect," he adds, "we can easily bring living animals into the lab for study. One technique is to use relaxing chemicals on the clams and scallops so that they open up under the microscope. A great deal can then be learned about how they gather food and move. Moreover, when the observations are finished, the animals can be revived and returned to where they were found. Movement can also be studied in the shallow water outside the lab. Scallops are fantastic swimmers, clapping their shells together like a set of false teeth biting through the water."

James Norris, another member of the research staff, goes off the reef's outer rim for some of his collecting: "You make a 50-foot descent through crystal-clear waters to get to the reef rim. The next step is like leaping off a mountain. You face a vertical drop of 600-700 feet, with giant sponges, marine algae, and 'forests' of sea fans on the steep slope. They pass you by as you go downward. It's a sensation of free-falling into seemingly bottomless depths, and you have to keep reminding yourself that there is a bottom in the dark blue."

A specialist on marine vegetation, Dr. Norris is interested in natural compounds contained in certain reef plants. As he explains it: "We think these compounds protect the plants from the grazing of invertebrate animals and fish, saying in effect, 'Don't eat me, I'm toxic.' " He collected a large iridescent blue-brown algae at Carrie Bow Cay that contains unique natural compounds and is potentially useful in pharmaceuticals, herbicides, and insecticides.

Plans to drill down into the Carrie Bow Cay reef itself to determine its history have been drawn by Walter Adey and his colleague, Macintyre. They envision a series of closely spaced holes drilled across the barrier reef, from the lagoon to the outer reef rim. The hollow drill extracts a column of rock, called the core. With the extracted cores radio-carbon dated and analyzed in content, the sequential history of the reef over thousands of years can be reconstructed.

Both scientists have been involved before in Caribbean reef drilling projects. Macintyre did his work off Panama with the aid of a newly designed submersible drill. "Unlike previous studies which have been restricted to shallow or exposed reef areas," he says, "I've been able to extend my drilling to deep reef zones and chart the relationship of the rising sea level to reef development after the melting of the last great ice sheets."

For several years Dr. Adey has worked in the Lesser Antilles off a trimarin sailboat, which serves double duty as a dwelling and a laboratory. Anchoring off a reef, he typically uses a rubber dinghy to transport his equipment to the reef crest where a drill platform is erected. If the bit doesn't get stuck in the coral bubble, he says, he can drill through a 40-foot-thick reef in a single day.

"Stony corals are the chief reef building agents in both the Caribbean and the Pacific when wave conditions are fairly quiet," Adey notes. "But when we drill in areas of high wave energy we find a record of reefs primarily built by calcareous algae. Reefs like these were once thought foreign to the Caribbean. We now see that this isn't

Amphipods, tiny planktonic crustaceans such as this, abound in the Carrie Bow Cay waters, where they are an important source of food for many of the larger reef animals.

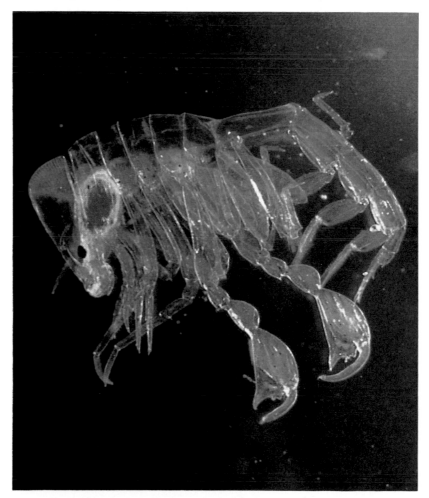

Two Islands

true. They are here and just as well developed as the Pacific reefs."

While calcareous algae can play the key role in building the reef, red, green, and blue-green algae are responsible for the reef's high productivity. This uncalcified algae, which grows like grass in the shallow reef waters, helps support the reef's animals by transforming sunlight and nutrients into food and oxygen.

"The importance of this algae lawn has been generally ignored," asserts Adey. "Perhaps it happened because most of the scientists who worked on reefs were zoologists. They watched fish and the other larger, 'more interesting' animals. But not the plants. Fortunately, it's different now. Because of the increasing interest of energy-conscious ecologists, the role of algae has become apparent. It explains something that once puzzled scientists: how you can get a highly productive reef in the middle of the sea, when the sea is generally very poor in nutrients. It is true that some plankton drifts into a reef area from outside. But my studies show that this only accounts for a small part of the high biological productivity that occurs right on the reef."

To duplicate conditions on a reef like Carrie Bow Cay's, Adey has built a 1,500-gallon aquarium tank system at the Museum with the light intensity of tropical noon, a night-day light cycle, and a wave machine. He chopped out whole chunks of reef rock, along with the surrounding and internal organisms and water, and shipped them to the Smithsonian. "Most people who set up a tropical aquarium bring in fish, maybe a crab and a few other isolated organisms," he explains. "But as soon as you isolate living beings from their natural ecosystem, you have to start caring for them in a very artificial way—by putting in food, filtering the water to eliminate their wastes, and so on.

"But we're trying to keep our animals and plants alive by providing the same environmental conditions that are found in tropical waters: constant warmth, a lot of light from overhead halide-vapour lamps, and water movement generated by a wave machine. The only 'extra' thing we're doing is throwing a little dried shrimp into the water once a day to simulate the plankton that enter a natural reef system daily," he continues. "That input is about balanced by removing excess algae growth, much as it is normally carried into the lagoon in a real reef. We have hundreds of species. Most are healthy, and our measurements show that their productivity is comparable to that of a tropical reef. With a larger tank and more light, we hope soon to create an environment where we can have thousands of different reef organisms living together, just as they do at Carrie Bow. This will make it possible for us to carefully monitor a reef population in a closed system for the first time. We'll be able to see interactions between the animals and plants on the reef that aren't seen while diving. For now, we have a living reef," he concludes. "As we increase its size and complexity, and learn how to keep the most delicate organisms, we should begin to understand the fabric of ecological relationships that sustain a complex reef community like Carrie Bow Cay."

IV. Looking at Culture

The most welcome words to a Smithsonian anthropologist in the field: "Come, sit by my fire and listen to stories of my people." This, of course, in a softly sibilant, exotic language.

Canela Festival Games

Central Brazil

The larger world is catching up with the Canela, Indian tribespeople of Brazil's central savannahs. Surrounded by increasing numbers of backwoods settlers, Canela economy and life style are finally being transformed. Smithsonian anthropologist William Crocker has spent long periods with the Canela, observing these changes.

"Happily, many Canela traditions are still intact," observes Dr. Crocker, "and among the most delightful are their festivals and games. Ceremonial days begin well before daylight. I awaken in my hammock on these occasions at 3:30 a.m., to the sounds of singing outside in the village plaza. Two to three dozen young women have formed a long line and are moving their knees and swinging their arms and singing. Before them dance 30 to 40 of the village's men, kicking up sand and stamping their feet to the rhythm of a leader's maraca rattle. The gaiety is infectious, even that early in the morning."

The Canela will amuse themselves almost around the clock by singing, dancing, and playing games on these holidays. Just before the sun rises, everybody goes to nearby streams to swim and bathe and cool off. When they return, the older men gather in council to decide how the day should be spent. Even if it is a workday they may choose to have a late afternoon log race—one of their favorite pastimes.

They use logs weighing as much as 300 pounds, especially cut for

Canela men trotting back to the village plaza after a masked festival, below. Sticks were used to support large, tent-like grass masks. Opposite, two children reflect the exuberance of the Canela.

Morning breaks over a Canela village, top. At right, men take off on an impromptu log race. The Canela run a variety of races of varying degrees of formality and organization.

The Magnificent Foragers

the particular contest. The tribe's youths begin training for these races when they are 12 to 14 years old. In earlier days marriage was not considered for a young man until he could handle the great logs, and today this ability is still regarded as a sign of maturity.

When the logs are large, four men will lift one and place it on a racer's left shoulder to start a race. Depending on how strong the starter is and how heavy the log, he will run along for 10 to 100 yards until he tires and transfers the log to the left shoulder of a teammate running behind him. Running a course of five to six miles, the team which gets its log back to the village first wins.

Great excitement is associated with these races. During the principal ones, village girls pour water from gourds onto the shoulders of their relatives who have just carried a log and need to be refreshed. The winning team, when it arrives at the village after a certain ceremonial race, sings triumphantly, trotting in front of the houses grouped around the circular village boulevard.

Afterward, team members bathe and refresh themselves in the streams before coming back for foot racing and later afternoon dancing in the plaza. Meanwhile the village elders gather for another meeting. When this concludes, everyone goes home for the evening meal. Another dance gets under way and continues until about 11 p.m. Then everybody goes home to sleep.

"Well, almost everybody," Crocker recalls, "for on such ceremonial occasions, a group of young men and women sometimes sing until well after midnight, walking very slowly around the village circle. These young entertainers are not carousing, they are bringing peace and tranquillity to others while enjoying themselves. Their aboriginal minstrel singing is carried out in a minor key with the chords sustained for 10 to 20 seconds before passing on to the next proper harmonic resolution. The certainty and precision of these beautiful musical pro-

Their ends carved out to provide hand holds, these puh-ruh logs might weigh as much as 125 kilos. A grueling relay in which the logs are passed from shoulder to shoulder, this particular race is one of the biggest held during high festivals.

Over the course of field work, William Crocker has become a trusted member of the community, participating in festivals but preferring his observer's role. Above, he is formally attired in falcon down, glued to the body with tree resin and coconut oil, for a ceremony. Opposite top, snaking along sideways in a modified conga line, young Canela dance through the village. Girl, below, is being made up for festivities. She is one of two girls who are special members of an old men's group, an important social association for the Canela.

gressions used to put me to sleep feeling the great confidence these troubadours had in the permanence of the social order of their society."

But change is at last looming for these people. Left free to develop in their own way for more than a century while other less isolated tribes in Brazil changed rapidly or vanished, the Canela now find the population of settlers around their reservation growing at a fast clip.

The Brazilian government has designated the state where the Canela reservation is located on aboriginal lands as a settlement area for independent farmers. In 15 years, when there may be 100 times as many Brazilians in the region as there are now, pressures are likely to reduce the size of the Canela reservation, requiring the tribe to give up its traditional slash-and-burn agriculture.

Canela agricultural traditions are already changing. The Canela have abandoned the potatoes and yams they once subsisted on in favor of the nutritionally inferior rice and bitter manioc eaten by the backlanders. Nor are the Canela able to depend on wild game and fruits as they once did for the principal share of their food supply and earnings. Wild game is becoming scarce, and while plenty of savannah berries and fruits are still available, the Canela don't eat them because the Brazilians, whom they now admire, hold this practice in low esteem, associating it with a state of savagery.

More and more the Canela rely on activities that approach begging. They go out in small family groups and live as squatters as far as 50 miles away, next to small ranching communities. Here they carry water, cut wood, and garden for Brazilians who will feed them.

The Magnificent Foragers

While there they pick up the Portuguese language, Roman Catholic religion, and many new values. They've even learned to dance the samba, according to Crocker.

"These changes, however, need not mean that the Canela culture is doomed," he surmises. "The Canela have survived since their military 'pacification' in 1814 because circumstances have permitted slow change—generation by generation. I find it consoling to hope that even if enormous population pressures do force rapid changes upon them, it could still be possible that bright and aware young Canela leaders might be able to cope, choosing what they like and avoiding what they abhor of the emerging backwoods culture around them. This could mean a relatively happy future for these fun-loving people, including a pleasing mixture of traditional song-dancing along with modern samba shuffling."

Looking at Culture

Collecting in the Forbidden Kingdom

Bhutan

Because of the accelerating rate of change in different cultures, the Smithsonian compiles contemporary folk collections that it regards as invaluable records of vanished, or fading, ways of life. Not long ago, Eugene Knez, a specialist on Asian anthropology, visited the small country of Bhutan, high in the Himalayas, to document its changing culture.

For centuries Bhutan maintained its feudal traditions and political independence by keeping its borders sealed to the outside world. Thus it gained its reputation as a "Forbidden Kingdom." Now it has ended its self-imposed seclusion, become a constitutional monarchy, and instituted economic, political, and social reforms. Modern schools, hydroelectric stations, hospitals, and a postal and telegraph service are in operation. Buses, trucks, and cars go back and forth daily on a new road connecting Bhutan and India. Even so, alongside these signs of 20th-century progress, the colorful character and individuality of old Bhutan are still in evidence.

Traveling through this steep-sided country, Dr. Knez was able to gather objects reflecting the mixture of ancient and modern life. "At all levels, religious character is one of the most striking aspects of Bhutan," he reports. "Buddhism is the state religion, and there are numerous major monasteries and small shrines. I collected objects that illustrate religious practices: a ceremonial mask, a Buddhist incense burner, a candlestick holder, and a temple gong. Individually, these items are modest in appearance, but taken together they say a lot."

To show how Bhutanese boys and girls are raised and educated, Knez made a collection of toys, books, and clothing. Besides tradi-

Bhutanese men from the northwest part of the country display traditional warrior garb, including armor covered with yak fur, iron and brass helmets, shields of leather and brass, and belts of silver and yak fur.

The Magnificent Foragers

The tanka, or scroll painting, above represents Padmasambhava, a Buddhist lama who brought Tantric Buddhism to Bhutan in the eighth century A.D. Padmasambhava is seated on a tiger skin because he was supposedly flying on the back of a tiger when he first arrived in Bhutan. Tankas painted by lamas were used as banners in religious ceremonies. The heavy modern boots, above right, collected by Eugene Knez in 1974, were worn by an official—or by a dancer depicting an official. Probably sewn by a Tibetan refugee shoemaker working in India, they consist of Indian rubber soles and leather toes, Bhutanese cloth uppers, split backs brocaded in black silk, and insteps of brocaded colored silks. Nepalese immigrants, below right, work on a recent Bhutanese roadbuilding project.

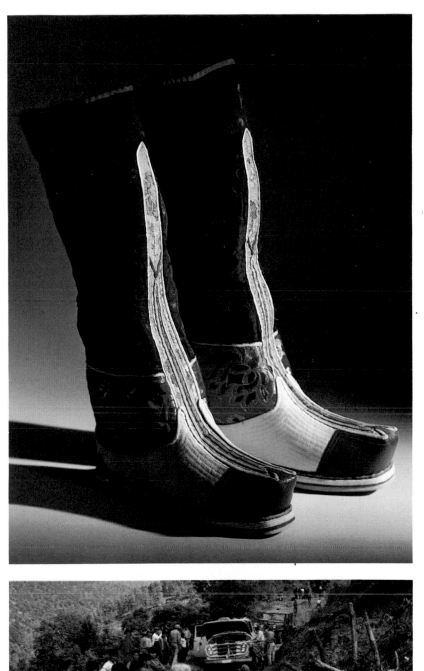

tional objects, he also collected manufactured items just coming into use there: a clock, for example, where before there were no clocks; or a pot made out of plastic, where previously all of the vessels were made from native materials.

"In a rural village," says Knez, "I would often attempt to create a time capsule of objects for that particular time and place. An object is, of course, only half of what is needed. Pertinent information is equally valuable. It's necessary to know how the object is used and how and of what kind of material it was made."

Handicrafts acquired by Knez include decorated copper and brass plates, gold and silver vessels, and samples of the brilliantly colored textiles woven in Bhutanese homes as a cottage industry. Women wear this material as a wraparound dress, held near the shoulders with silver brooches and chains. Men favor a long, colorful coat tied around the waist with a sash to keep it knee-length. Bhutanese often go barelegged and barefooted; the more prosperous wear woolen Tibetan-style boots, or Western shoes with long stockings. Basketry used in daily life is varied. Split-cane containers are made to hold cooked foods and are treated with resins to hold water.

Americans tend to think of Asia as crowded with hordes of people, but, according to Knez, traveling in Bhutan is somewhat like being in the American West. "After you pass one house you can go a long way before seeing another," he says. Often located high on mountainsides, the homes are usually three stories high and made of large timbers. The family lives on the second floor, using the third floor for firewood and storage and the ground floor for its livestock.

"Occasionally I would be invited inside one of these country homes," he recalls. "I was interested in acquiring agricultural tools, and when I saw one that I wanted, I would ask if they had an extra one which I might buy. Soon a series of tools would be on the ground in front of me, and the bargaining would commence. In the end, everyone would be satisfied with a sale. The family now had some extra cash and could replace, if desired, the items they had sold, and I had obtained used artifacts in good condition—items dear to a field collector's heart."

The woman and baby above, like the majority of Bhutanese, are Bhote, a tribe of Tibetan origin who tend to be Bhutan's farmers and landholders. The house, above right, displays a typical Bhutanese roof of handmade wood shingles held in place with rocks. Masks such as the brightly colored papier-maché one on the opposite page are used in religious dance dramas. The third eye in the center represents the power to see into the future; the skulls represent deceased Buddhas.

The Magnificent Foragers

Hairdress Among the Himba

Southwest Africa

Driving their cows through the desolate, hilly country of the western Angola-Namibia borderland, the Himba move their encampments whenever water and forage conditions dictate. Their homesteads and cattle posts lie sparsely scattered in the mountainous basin of the lower Cunene, one of the few rivers that cross the desert stretching for hundreds of miles along southwestern Africa's coast.

Over the years, the Himba have been visited periodically by Smithsonian anthropologist Gordon Gibson. Camping close to Himba homesteads, he collected information from the residents with the help of an interpreter, a tape recorder, and a camera. Generally his interest stimulated a good deal of rapport. "They knew that I was not there to tax or innoculate them," he says, "and they like talking about their lives and the tribe's history."

Dr. Gibson has photographed and collected voluminous notes on

Himba herdsmen often lead their cattle far from the family encampment in search of water and grass or browse. Armed with a short spear and club, they are prepared to protect the herd from lions and other marauders that roam the arid wilderness in which the Himba dwell.

Richly smeared with red ochre and butterfat, Himba women prepare one another's coiffure. The strings twisted at the back of the head are lengthened with hair shaved from the forehead or taken from the queue the husband gave up at marriage. Ash and powdered herbs added in the process make the strings pungent and stiff.

the daily life and social customs of the Himba as well as on the ceremonies and feasts with which they celebrate births, puberty, marriages, and funerals. He has also recorded the epic songs that the herdsmen sing of raids and battles with neighboring tribes in bygone days.

"But I am equally fascinated by the artistry that the Himba exercise in personal adornment," says Gibson. All of the people in that part of Africa wear outstanding headdresses. When he first saw the Himba variety, he set himself the task of unraveling their significance. He found that their striking styles of headdress and coiffure and their use of red cosmetic powder and pungent herbal perfume distinguish them from neighboring tribes. Later he discovered that they serve as symbols of a person's sex, stage of life, clan membership, and status.

Much time and effort is devoted to these adornments. During the middle of the day, when the chores of watering the cattle and working the fields are done, the Himba occasionally sit in the shade and dress one another's hair. "It's sort of an open-air beauty parlor," comments Gibson. The hairdresser has two containers—one holding ash, the other powdered herbs—into which he or she dips from time to time as needed. Both sexes add dried pungent herbs to their hair. Women perfume their clothing with the same powder, burning it like incense

Looking at Culture

and suspending their clothing over the smoke on conical wickerwork frames. In the privacy of their homes they perfume their bodies by crouching over the burning incense.

Though tradition sets the major styles of headdress and hairdo, Himba hairdressers display individual skill and imagination through variations on the basic themes. Curiously enough in this torrid country, tradition decrees some form of head covering for married people, except those in mourning. Women, especially, consider it indecent to go about with the head not covered, though the body may be scantily clothed. For ceremonial occasions the wives or sweethearts smear a mixture of powdered herbs, red ochre, and butterfat on themselves and on their men. Usually the salve is applied only around a man's neck, whereas a woman will have it spread over her whole body.

Boys and girls pass through an established succession of hair styles and ornamentation which grow more elaborate up until marriage and are finally simplified, at least for men in old age. The hair of infants is shaved off, with perhaps a little ridge left along the crowns of boys, or two tufts remaining over the temples of girls. Older children wear their hair in a variety of patterns, each distinctive of the patrilineal clan to which the child and his father belong.

As boys approach puberty, the hair is allowed to grow long on the top and in back, so it can be braided into a queue, while sides of the head are kept shaved. The queue is the mark of a young man who has passed through the circumcision ceremony but is not yet married. Girls wear their hair divided into two braided queues that hang down on either side of the face, and the hair in back may also be braided. At puberty a girl's forehead is shaved and the hair in back twisted into innumerable strings, each of which may be lengthened by the addition of plant fibers. A wood-ash treatment keeps the twists in place, and the mixture of butter and powdered herbs makes them shiny and fragrant. On reaching puberty the Himba girl wears a veil of strings of beads and seeds. It covers her face if allowed to drape naturally, but after a period of seclusion she often pulls the strings aside to show her face. When she marries, the young woman replaces the veil with the *ekori,* an elaborate sheep- and calfskin bonnet decorated with iron beads and bearing three short ears that stand up in back. It appears only for special occasions, such as life-crisis rites. Married women wear a ruffled cap of goatskin for everyday use. Girls married as children are also privileged to wear the beaded veil and headdress.

When a man approaches marriage—usually in his mid-twenties—his queue is divided into two strands, braided, and worn thus until the wedding. Afterward he allows the hair to grow, coiling the braids on top of his head and covering them with a skin or a cloth. Eventually the queues are cut off and given to his bride so that she may incorporate them into her twists of hair, to lengthen them.

The Himba subsist largely on the meat and milk of their herds of domestic animals, using the hides for clothing. Sheep- and goatskins are softened by rubbing them with butterfat. If pieces need to be joined, the women sew them together with sinew and awl. Men wear simple aprons of softened skin draped over a tubular leather belt. Women wear a whole animal skin as an apron, tying the legs around the waist. For ceremonial dress they wear sleeveless cloaks of skin ornamented with medallions of iron beads. Both men and women wear

Special hair styles distinguish the sexes and the patrilineal clans to which Himba children belong.

leather sandals, though soles cut from pieces of discarded automobile tires are popular when available.

Both men and women prize highly as ornaments the conical shells of coastal mollusks, the men wearing the circular end of the shell, and the women wearing whole shells on necklaces decorated also with iron beads. Iron and brass bangles made by local craftsmen are also worn by both sexes.

The tribe has lived along the Cunene River for hundreds of years, left alone for the most part because the dryness of the area made it unattractive for settlement by cultivators. Now this bucolic existence is threatened, for the bitter Angolan Civil War has made the Angola border a region of hostilities, and the future of the tribe uncertain. No one knows for how long the Himba will be able to pass their afternoons grooming in peace and tranquillity.

Looking at Culture

The Navigators of Puluwat Atoll

Caroline Islands

Puluwat is one of the last islands where the great voyaging tradition of ancient Pacific mariners is still carried on. The seagoing canoes that the Puluwatans build and sail so skillfully give a heroic quality to their society and sustain its vitality: they enable the islanders to fish on the high seas and to trade, intermarry, and promote political ties with their neighbors in the central Caroline Islands. Boats are built for all purposes on Puluwat, from small paddling canoes for commuting around the island to large outrigger sailing canoes for ocean trips.

The mariners who sail these remarkable boats practice vanishing arts of navigation dating back to when their seafaring ancestors first peopled the Caroline Islands of the Western Pacific. Because their ancient navigational knowledge could become extinct if not written down, Saul H. Riesenberg, a Smithsonian anthropologist specializing in Pacific studies, spent months questioning the captains of Puluwat's seagoing craft about the vast body of sea lore they command.

Puluwat navigators steer by the stars and the sun. They locate islands by the shape of a cloud formation or the flight of birds, and they hold a course by the feel of waves and swells hitting the boat's side. Compasses, of recent introduction, are relied on to check the course only now and then, to make sure that it is correctly reckoned.

Today, when large seagoing outrigger canoes sail out of Puluwat's picturesque lagoon, they are usually bound either for favorite fishing spots on the nearby reefs or for one or more neighboring islands. If they visit an island 150 to 500 miles away, they often linger for weeks to trade and enjoy the hospitality of acquaintances or relatives before returning. "The men of Puluwat have a kind of machismo about sailing," says Riesenberg. "Often they'll pick up and set sail for another island the same way we might impulsively decide to get into our car and go to a movie. They'll be sitting around partying, and one of the navigators will suddenly say, 'I think I'll go to Pikelot. Come on, who will come with me?' At that, three or four persons will get into a boat in the dead of night and set sail. Pikelot Island is about a hundred miles away, a 36-hour round trip. It's a favorite place because large green turtles can be caught there and brought back for big feasts."

Navigators command enormous prestige on Puluwat, and most of the island's young men aspire to join the elite. According to Riesenberg, very few make it. The training is difficult and demanding. An aspirant must memorize and employ an enormous amount of complicated and arcane information. Only about a half-dozen men on the island are recognized as master navigators. These seasoned mariners instruct younger men in the arts of navigation, free of charge if the student is from the instructor's family, for a price if he is not.

Instead of modern navigational aids, Puluwat navigators employ a wondrous trove of mental imagery. For the captain of a seagoing canoe in the central Carolines, the ocean's surface is studded with hundreds, perhaps thousands of elements whose names, locations, and relationships he knows and which function for him as sea marks. Some of them are real enough natural phenomena; some are in the realm of the mythical. For a scientist like Riesenberg, they are both a joy and a fascination to collect and record.

The Magnificent Foragers

"For instance," he says, "the islets, the reefs, and the sandbanks are certainly real, as perhaps is the discolored patch of water or the shark that is always to be found at a particular reef. But then," he adds, "there are the two-headed whales, the hovering frigate bird with the plover constantly flying circles around it, the spirit who lives in a flame, the man in a canoe made of ferns, and the ghost islands that do not exist—at least not physically.

"All of them, however—the fantastic monsters, the ghost islands and the real islands alike—most certainly do exist for the navigator as permanent geographical features. Though they are described metaphorically, all of them are more or less useful sea marks, all of them bear names, and all of them are taken quite seriously as navigational aids—some of which are kept closely guarded secrets."

Navigators arrange these reference points in such a way as to organize the miles of water into a schematic arrangement based on metaphor. For example, one of the systems makes use of the image of six whales in a row. The imaginary leviathans line up east and west with their heads to the north, like the skids used to launch a canoe. Each whale is positioned about one day's sail directly south of an island. Setting a star course, the navigator sails in his imagination to the first whale, turns north to the island south of which it lies, turns back south again to the whale, then west to the second whale, west to the third whale, and so on, until he comes to his destination.

In other systems the navigator pursues an ever-escaping fish while seeking to spear it at esoterically named reef-holes on successive islands; or he seeks a fish with the aid of a torch as he hunts it from place to place. The figure of speech in each case is an instrument which enables the Puluwatans to learn and remember the relationships between geographical phenomena, and between such phenomena and the courses of the stars.

For the navigators, there is another advantage: the more confusing and pedantic their art, the more impressive and inaccessible it is to other Puluwatans, enhancing the prestige and social standing of the elite to which the possessor belongs.

A canoe leaves Puluwat lagoon headed for the open sea.

Looking at Culture

V.   Plants of the Kingdom

In search of all that flowers: plants feed us, give us oxygen to
breathe, soothe our eyes, and lead intricate—sometimes
secretive—lives of their own.

Orchids

Ghana

To most of us, orchids—perhaps the most ceremonial flower in America, at least at banquets—are as placid as they are beautiful, their petals, shape, and hue seemingly the result of art as much as nature. But to botanist Edward S. Ayensu, who travels to the world's tropical regions to study their anatomy and pollination, orchids are an intricate, highly evolved, biologically fascinating family.

Take their system of seed dispersal and germination, for example. Each flower produces a capsule with thousands of microscopic seeds that are scattered from place to place by the wind. Ordinary enough. But after they reach the ground, the seeds must contact a "nurse" fungus before they can germinate, which is probably fortunate. For if they were not limited by this natural biological control, orchids would proliferate so fast that they would soon cover the face of the earth. As it is, they are still the most successful and largest of the plant families. Between 30 and 35 thousand different kinds are known, and new species continue to be discovered.

Some of Dr. Ayensu's explorations take place in his native Ghana and the West African tropical rain forest. In the field, he cruises the lakes and streams in a dugout canoe, scanning the banks for orchids growing on trees or shrubbery. When he sights a specimen like the spectacular, yellowish *Ansellia* clinging to a tree high above the ground, he first makes certain that enough other individuals of its kind are growing nearby.

Since many people in the forest regions of Ghana believe unusually large trees hold divine powers, Ayensu sometimes takes a priest along so the people will not think he is incurring the displeasure of the gods. When he finds such an imposing tree bearing an interesting orchid, therefore, collecting is put off until the priest pours a libation of schnapps on the ground and recites incantations asking the gods for blessing and protection. Only afterward do Ayensu and his field assistants bring down the orchid with the aid of a long bamboo pole.

African orchids, such as Ansellia africana, *right, have been studied less intensively than their New World counterparts for the simple reason that they are less flamboyant. In an evolutionary sense, however, they are equally sophisticated.*

Plants of the Kingdom

Birds, bats, bees, moths, butterflies, wasps, and other pollinating insects have evolved elegantly adjusted partnerships with tropical orchids. The flowers produce nectar, thus feeding these creatures and in turn assuring their own pollination, fertilization, and seed production. Ayensu finds the workings of this interaction especially intriguing.

"Orchid flowers are structured to attract the specific pollinators essential to reproduction," he says. "In many cases orchid flowers seem custom-tailored to fit the bill of a hummingbird or body of a bee. Another design factor is the color of the flower. Studies have shown that bees are drawn to violet, blue, green, and yellow, colors at the ultraviolet end of the spectrum. Orchids and other flowers have developed blossoms in these colors. Fragrances emitted by the orchids also play a key role in their pollination. Euglossine bees, for example, are attracted to particular scents given off by different orchid species. The odors are produced by as many as 39 chemical compounds within the flower—some of which smell like women's perfumes."

Attracted to the lip of the blossom by color, the bee alights and, in search of nectar, enters the flower's inner chamber. As it leaves the flower, its body rubs against the stamen of the flower and picks up sticky packets of pollen grains. Moving on to the next blossom for another intake of nectar, the bee brushes these pollen grains onto the next flower's pistil and thereby cross-pollinates it. "It is a well-known process but still an exquisite, sophisticated, and profound one," Ayensu acknowledges.

Botanist Edward Ayensu, seated with pipe, relaxes with villagers in a small town in Ghana, near the eastern border with Togo. Collecting orchids in the field often requires a good working rapport with local residents and an understanding and respect of their customs.

The Magnificent Foragers

Unusually colorful, this Renanthera storiei *is one of the most common of the African cultivated orchids. A terrestrial orchid, it is frequently collected by Africans for their gardens. Like most African orchids, it has relatively small blooms which grow in rows on a stalk, or bract.*

One of his projects involves a time-lapse photographic study of how orchids synchronize the emission of their fragrances with the opening of their blossoms. Going into the field, he selects a group of flowers beginning to open. Setting his camera on a tripod, he photographs the flower once every minute or every three minutes, depending on the speed with which the flower opens. Some open quickly, others take two or three hours. During that time, bees tend to home-in on the flower since the production of sweet fragrances is at its peak. Ayensu monitors the fragrance production from minute to minute with a pocket spectrometer, an instrument that measures the flower's nectar concentration.

Plants of the Kingdom

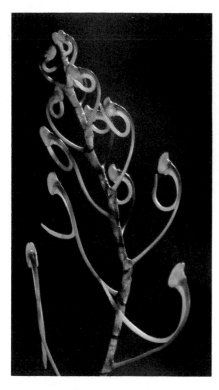

But that occurs during daylight. Some orchids open their petals at night to receive moths and other nocturnal pollinators. To capture these events on film, as well as to study the nocturnal behavior of bats, pollinators of tropical blossoms, Ayensu attaches to his camera an "Owl Eye," a portable instrument that amplifies available light an average of 20,000 times.

Photography has proved an important adjunct to Ayensu's studies in the laboratory as well as in the field. He has constructed an optical shuttle system at the Smithsonian to observe the internal architecture of orchids and other tropical plants. He photographs thousands of transparently thin stem cross-sections, and then sequentially superimposes the images over one another, producing a motion picture that follows the movement of the plant fluids as they flow along the interior ways of the stem.

"I believe most people think that a plant like an orchid is very static," he says. "But my studies attest to the fact that there is much dynamism and movement in its life; not the same as an animal's, of course, but containing great vibrance nevertheless."

In spite of their amazing abundance, orchids are in danger. The wilderness habitats that sustain them face increasing pressure from the exploding world population, and in various countries this problem is aggravated by commercial exploitation. Some rarer tropical orchids sell for hundreds of dollars and are heavily collected by garden enthusiasts and commercial exporters. Many of the 211 kinds of orchids native to America, for example, are either rare, endangered, or threatened due to the destruction of their forest habitats. Ayensu's office at the Smithsonian lists and monitors the status of all United States plants in this predicament, an alarmingly high number. Some

Three different African orchids: above, an Angraecopsis gracillima; *below, a* Diaphananthe rutilla; *right, an* Angraecum eichlerianum.

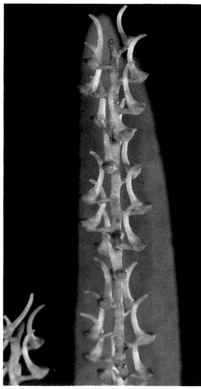

The dog-faced bat, Epomorphorus gambianus, *surprised at rest. Fructivorous, or fruit eating,* Epomorphorus *is botanically interesting because it is a primary pollinator of fruit trees.*

Designed originally for night maneuvers by the military, the "Owl Eye" system was adapted by Dr. Ayensu for non-intrusive night photography of orchids and bats, particularly feeding and pollination activity. The Owl Eye amplifies available light about 20,000 times, producing exceptionally high-resolution images. Unlike infrared systems, this equipment does not emit any heat or light, which startles bats.

For all its delicacy, this Spathoglottis plicata *is a mover. An escaped import, it is native to the Caroline Islands in the Pacific but now flourishes in Hawaii and even in Africa. In Hawaii, apparently, its seeds impregnated the soil around imported reforestation tree seedlings, quickly germinating and establishing the plant on the islands.*

2,000 species are currently in jeopardy, almost 10 percent of the total native flora of the continental United States. Half of the Hawaiian flora is similarly endangered.

"Most of our concern about endangered wildlife has been directed toward animals," he points out. "Now, for the first time, we are beginning to understand that plants are also in trouble. Some plants are dying natural deaths, while some will become extinct no matter what we do; but man's activities will certainly trigger the deaths of many vital species unless we take immediate steps. We must not forget that these plants are the life-support system for all organisms on earth, including man.

"Many plants, like orchids, are too beautiful for their own good," notes Ayensu. "Cacti, succulents, lilies, and carnivorous plants are also seriously depleted. Cacti are being dug up by the truckload in desert areas and sold in plant shops. And, like tropical orchids, the greatest pressure is on the rarest species because they command the highest prices. Some species of cactus and orchids are now so rare that they could be rendered extinct in minutes. Literally in minutes."

The Magnificent Foragers

In Search of Native Flora

Lush Hawaii, with magenta bougainvillea in the foreground. A surprising number of the islands' plants are imports, many brought by early settlers.

Hawaii

Since Captain James Cook's voyages of exploration in the 1770s, profound changes have taken place in the natural environments of many of the Pacific's small and isolated islands. Cattle, pigs, and goats—all subsequently introduced—overgrazed and trampled these virgin landscapes. Man's other works created havoc as well, and continue to, as land is cleared for agriculture, roads reach into all parts of the islands, tourists travel about, and the human population grows and grows and grows. One result of such constant pressure is that many kinds of plants can't survive it. Smithsonian botanist Raymond Fosberg laments this state of affairs, exacerbated in his view because far-reaching changes in the flora often go unnoticed through a lack of authoritative botanical data.

"We've been collecting plants throughout Oceania for more than 200 years, and you'd think that everything would be known," says Dr. Fosberg. "But just try to get a complete list of the plants of most of the islands. You'll find you can't. For as many species that are known, there may be just as many that aren't. For every bit of botanical information that we fail to get on record, we decrease our own chances of survival by that much. Things are changing so rapidly that if we are ever to get a well-rounded body of information about the earth we live on—the conditions to which we are genetically fitted—we had better do it fast."

A specialist on the ecology of tropical atolls, Fosberg has visited Pacific islands since the early 1930s. The ones he knows best are the Hawaiian islands. When the Smithsonian began its United States endangered species list in 1974, he calculated that 1,750 kinds of Hawaiian plants were either endangered or threatened, a figure he regards as fairly typical for all of the Pacific's high islands.

Of the high islands—those that rise far enough above sea level to make possible a varied topography and diverse plant population—almost all have been influenced by man in a destructive way. One of the most harmful developments has been the introduction of four-footed animals like goats, cows, and sheep. Since the plants had never contended with such grazing animals before, they never developed any protective immunity—no thorns, poisons, or bad taste. Thus, when these domestic animals were turned loose, they were able to gorge themselves on almost every plant that grew.

"This would be followed by soil erosion," Fosberg points out. "Fly over some of the smaller Hawaiian islands and you'll see a plume of red mud spreading in the ocean for a mile downwind. Islands, of course, are small places, and the plants on them grow close together in small populations that are unique, having evolved in that special spot for millions of years. Wipe out a population of plants on a small island's little mountain, and the species becomes extinct; whereas on a large continent, you'd probably find the same plant occurring on the other mountaintops."

Once these unique plants are extirpated, their place is taken by common weeds, often carried by man. According to Fosberg, such artificial succession tends to produce uniformity. "You know," he says, "if you go to Hawaii and are not a trained botanist, you could stay

there for as long as three months and not see a native plant. You'll see mesquite trees, hibiscus from Florida, and weeds or grasses common to the District of Columbia. You'll see guava trees, mangoes, and all manner of plants that are widespread in the tropics. But, unless you are taken to less disturbed places, you won't see any native plants."

Fosberg and a colleague, Marie Sachet, who works with him on Pacific Island ecology, offer two practical reasons for saving these surviving native plants. "First," he says, "it is impossible to pick out a certain set of resources and say that these are going to be the ones that will always be desirable. We can't predict the future. If we could, we could select out the plants which will never be important and then let them disappear. Second, it is essential to save every possible kind of plant on the chance that it may be a source of resistance to plant diseases. It is well known that ordinary cultivated plants have varying degrees of resistance to disease. Plant blights are relatively short-lived episodes that appear rapidly and may suddenly devastate crop plants.

"You can stop some diseases with chemicals," he continues, "but most have to be combated by breeding plant varieties that are resistant to the diseases. What you do is try to get genetic material from all over the world—both wild and cultivated—and cross many varieties until you get new seed stocks that can't be infected. Without a great assortment of material to experiment with, you may not succeed."

Finally, Fosberg thinks we need to increase our array of crop plants. Over thousands of years our ancestors were able to domesticate a great many plants, mostly by trial-and-error methods. But in the last 100 years, even with a sophisticated knowledge of breeding and genetics, only a relatively few new plant varieties have been developed. It takes a great deal of time and patience to do so. Still, Fosberg believes there exists real potential for domesticating thousands of other plants for food. "I think," he says, "we're now coming to the point where to increase our set of resources and keep ourselves from starving, we're going to need to do this."

Bamboo

Manaus, Brazil

Not long ago, a call came in to the Smithsonian for two botanists, Cleofé Calderón and Thomas Soderstrom, who interest themselves in bamboo, a giant grass that thrives in the tropics. Its pliancy, toughness, and tubular form make it such an excellent construction material that South Americans are looking increasingly at its economic promise. When an aerial exploration of the Amazon revealed enormous stands of native bamboo in Acre, Brazil's westernmost state, the Brazilian government called Drs. Calderón and Soderstrom to help evaluate this potential resource.

Radar-photo image equipment carried by jet planes is being used to map the Brazilian wilderness—a region two-thirds the size of the continental United States. Penetrating the thick clouds and the forest's heavily foliaged canopy, the apparatus makes it possible to discern different levels of terrain, possible mineral deposits, character of vegetation, and hidden streams and forests. Not surprisingly, more has been learned about the Amazon by this aerial technology than by four centuries of exploration undertaken previously on land and water.

After flying to a remote Air Force base near the Peruvian border, Calderón and Soderstrom made daily helicopter hops to various forest sites. "In many cases," Soderstrom reports, "teams of workers equipped with electric saws descended the ship on a rope ladder, then hacked out a clearing so the copter could land. Some areas had obviously never been visited by humans before."

When they disembarked, the botanists found thickets of spiny bamboo canes intertwined with forest trees, the bamboo sometimes arching as high as 100 feet and reaching the tops of the trees. While the helicopter stood by, they made collections of canes, foliage, and rhizomes for laboratory analysis. Plants and terrain were photographed in color and black and white. Notes were taken of maximum and minimum forest temperatures, soil and light quality, and other ecological factors. Specimens were shipped to the National Institute of Amazonian Research, a large Brazilian laboratory complex at Manaus on the Amazon River, where the fibers were tested to determine their potential for paper making.

Specimens also came to the Smithsonian for study. The Institution maintains one of the world's great bamboo research collections, brought from China and other parts of Asia and tropical America by the late F. A. McClure. He was fascinated by the observation that some species flower and die in cycles that sometimes do not reoccur for decades, or even for more than a hundred years. "It's a very difficult matter to study," Dr. McClure once said. "Life is too short for a man to concentrate on the problem long enough to solve it."

Ironically, McClure was to die at age 72 during a major worldwide outbreak of a long-term bamboo flowering. This one occurs at intervals of 120 years, as rare to botanists as Halley's Comet is to astronomers. After flowering, bamboo clumps die back. As many as 15 years may pass before a bamboo clump returns to normal. In Japan, where the stricken species of bamboo is used as a raw material for the construction of homes, furniture, farm implements, and baskets, the cyclical flowering was nothing less than an economic disaster.

New bamboo shoots grow skyward through the tall trees of a Brazilian forest. Once they have attained maximum height, they develop lateral branches that reach like arms over neighboring trees for support.

Plants of the Kingdom

The Giant Thorny Bamboo of India grows for about 30 years, producing a magnificent clump of tall, green-plumed canes, but no flowers. Then, suddenly, all clumps begin to lose their leaves, and flowers emerge at every node. It is a once-in-a-lifetime flowering, however. Thereafter, the plant dies, though its seeds begin a new generation of plants that will re-peat the cycle and flower again, 30 years or so later.

"Building on McClure's legacy," says Soderstrom, "we continued his study of this dramatic botanical event. Readers of *Smithsonian* magazine were asked to be on the alert for stands of flowering bamboo, and the same request went out to high school science classes across the country. The response was overwhelming. Mailbags full of dried spec-imens of flowering bamboo plants, and accompanying documentation, arrived at the Smithsonian. It soon became clear that the flowering was not taking place simultaneously throughout the United States. In-stead, it was progressive, involving one stand after another."

According to Soderstrom, a new time-oriented hereditary line originates whenever seed from plants on one hereditary line germi-nates in different years. Over a long time, the effect is cumulative, giving rise eventually to a whole series of distinct hereditary lines, each flowering at a different time.

What triggers these flowerings? "Although there are many theories," replies Soderstrom, "the pertinent questions are unanswered and still a challenge to the experimental plant physiologist. Fortu-nately, the material we have assembled will be valuable to scholars studying such clock-like biological phenomena."

The Magnificent Foragers

Endangered Palms

Jamaica

Among the earliest flowering plants to evolve, palms developed slowly over millions of years in relatively undisturbed tropical surroundings. Now, however, blights, deforestation, and other forms of environmental degradation are sweeping through the lands where palms are indigenous, threatening the survival of many forms of these showy, sensitive plants.

In the West Indies and Central America, where the Smithsonian's Robert Read conducts his field work, small, fragile populations of wild palms are being wiped out by logging and burning, and a currently unstoppable plant disease called "lethal yellowing," which has killed off thousands of cultivated coconut palms and a number of other exotic palm species.

"I worry about the disease jumping to the South Pacific and Southeast Asia," says Dr. Read. To understand why the thought makes him shudder, it is necessary to appreciate the importance of palms in the everyday economy of the people of Southeast Asia and Oceania, where palm products are a source of food, fiber, and construction materials. In Southeast Asia alone, millions rely for their

A fragment of a fan-leaf thrinax, or broom palm, of the West Indies, above. Below, a row of coconut trees reflects in a lagoon at Cape Coast, Ghana.

Plants of the Kingdom

The inflorescence of Cuba's Coper-nicia macroglossa, or petticoat-palm, above, contains thousands of tiny flowers and projects above the leaf. Right, last stages of lethal yellowing in a Jamaican coconut plantation. Bananas have been planted to stave off economic disaster.

own use on the coconut and other palms which provide them oil, sugar, a fermented beverage called toddy, alcohol, fibers, and thatch. But they also export many parts of this versatile plant to developed countries. Palm oil is sold for use in margarine and soap; coconut meat for a variety of bakery and ice cream products; leaves, fibers, and wood for mats, brushes, baskets, and rattan furniture; and hearts of palm and dates for table consumption.

Read's interest in palms led him to Florida and Jamaica, where lethal yellowing runs rampant. While in Jamaica he worked at determining the effect of the disease on different palm species, particularly on the less hardy exotic palms. Officials had already determined that hardier varieties of coconut will grow, which may lead the way to the recovery of the coconut industry, not only in Jamaica but elsewhere in the Caribbean.

The Jamaican experience may also help other tropical lands around the world protect their palm resources. It won't be easy, however, since a wide variety of diseases affect a wide variety of species. In North Africa, for example, a lethal fungus disease known as bayoud has for 70 years marched eastward unchecked across Morocco and

Algeria, striking the most desirable date palms and turning classic oases into barren, shadeless counterparts of the surrounding desert. As in Jamaica, the only hope lies in finding or developing a resistant date variety. Unfortunately for European connoisseurs, such species identified thus far bear inferior, poor-tasting dates.

Another disease known as hartrot or fatal wilt is threatening the coconut and palm oil industries in Surinam. Even if the lethal yellowing-resistant Malayan variety of coconut succeeds in rejuvenating the coconut industry in Jamaica, the threat of hartrot remains. The Malayan coconut is not resistant to the ravages of this disease. Furthermore, its symptoms have been observed in a number of wild palms, which could, as was the case with lethal yellowing, foretell spreading to the American tropics, or possibly even to the South Pacific and Southeast Asia.

The sheer numbers of palms present problems to botanists: more than 2,700 different palm tree types grow throughout the world. Different species range in height from a mere 6 inches to a towering 300 feet. Their trunks can be green and shiny, or gray and waxy; as slender as a pencil, or as thick as a large barrel. Leaves also vary greatly in size and can take the form of a feather, a fan, or a combination of the two. Some palms produce only a few flowers and fruits while others display a million or more on a single flower cluster.

Ponderous and complex, many palm inflorescences and leaves—some of them are the largest leaves in the plant kingdom—cannot be handled in the same manner as other plants. As a result, botanists have always had a hard time collecting and preserving them.

"In the old days," says Read, "collectors often selected the smallest leaves, or chopped up larger ones to fit them into herbarium

An inflorescence of Calyptronoma occidentalis, *top, the "long thatch" of Jamaica. The hundreds of tiny white flowers fall off in the form of tiny dunce caps, or "calyptras," hence the name of the genus. The tightly packed inflorescence branches and flower buds of a* Pinanga kuhlii, *a clustering palm of Java and Sumatra, immediate right. This species thrives in warm, moist botanical gardens, where it propagates readily from seed. Far right,* Caryota aequatorialis, *a tall, solitary fishtail palm, native to Malaya. This palm grows to about 15 meters in height, only to commence flowering from the top after the last leaf is produced. Inflorescences are then produced sequentially toward the base until the entire plant dies, some four years after producing the first flower.*

Plants of the Kingdom

A man climbs a coconut palm in Sri Lanka, top, to pick a special seed whose husk exhibits a pink color when sliced. Bottom, Corypha umbraculifera, *a fruiting specimen of the Talipot palm of southern Asia. Its fan-shaped leaves often measure 5 meters in diameter, and the inflorescence may spread 10 meters across and upwards.*

storage cabinets. That presents current researchers like myself with many problems. What can you say about a plant species on the basis of a single, incomplete, dead, fallen leaf-blade fragment? It's like trying to visualize an elephant from its trunk or its tail. In recent years I think we've improved our methods for studying palms.

"What I do," he goes on, "is study these awkward, and often viciously spiny, 'dinosaurs' of the plant kingdom in their natural environment. As I work out the parameters of variation, distribution, and ecology, I select representative specimens, take numerous photographs, and preserve selected samples of leaves, flowers, and fruits for later, closer analysis. At the same time fresh pollen may be collected to be germinated on nutrient media for chromosome studies."

Botanist Read, like those before him, searches through the jungles and rain forests of tropical America for various species of palms. Read, however, is searching for the missing pieces of palm puzzles, trying to find answers to as many questions as possible before disease and the woodcutter take their toll. Then, back in the Museum laboratory, he dissects leaf samples and germinates palm pollen in order to put the pieces of the puzzle together, thereby enabling him both to understand differences between palms better, and to clarify relationships among the many diverse palm species.

The Magnificent Foragers

Two Deserts

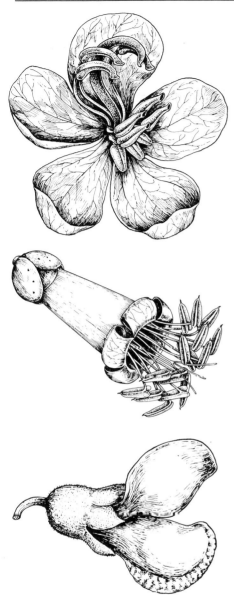

Mexico; Argentina

At first glimpse, North America's Sonoran Desert, with its tall, barreled saguaro cacti and its spiny trees, resembles South America's Monte Desert. But Smithsonian botanist Beryl B. Simpson suspected that more was going on than met the eye. She decided to compare the two supposedly similar areas.

Dr. Simpson's close examination proved that the vegetation of these widely separated areas with seemingly comparable rainfall, temperature, soil, and other conditions, is quite different indeed. Her studies carried her from arid Sonoran borderlands of Mexico and the United States to Argentina's dry Monte region.

A specialist in plant breeding systems, Simpson focused on the 13 most prevalent plants in each desert. She set out to determine what relationships these plants have with pollinators; how much energy the plants apportion to nectar and flower production; and how many pollinators are supported by the plant community. Drawing her conclusions from measurements of flower and nectar production, sugar concentration of the nectar, and amounts of pollen produced, she developed collateral data on blooming schedules, floral structure, and odor.

Her studies showed that plants of the two deserts are influenced by different weather patterns, resulting in the use of different pollinators. Plants in each region vary in the amounts of energy they invest in flower production, and they supply differing amounts of nectar and pollen to potential animal pollinators.

The blooming patterns of the two areas are also dissimilar. This may have been caused by the way in which rainfall is dispersed during the year. The yearly amount of rain at both sites appears to be about the same—10 inches—but at the Monte Desert site it rains only in the summer, whereas at the Sonoran Desert site in southern Arizona, rainfall, though unpredictable, usually falls both during the winter and briefly at the end of summer. As a result, in Argentina the desert plants tend to bloom patchily for a long period during the summer. In Arizona, on the other hand, most of the perennial species bloom in the early spring when temperatures begin to rise and again at the end of the summer if additional rain falls. In times of plentiful winter rain, a spectacularly colorful array of annual wildflowers also appears in the early spring.

Sonoran species that bloom in the earliest part of spring—before temperatures become excessively warm—are typically wind-pollinated. But not always. A characteristic Sonoran plant with brilliant large red flowers, the ocotillo, is pollinated by migratory hummingbirds on their way from Mexico to California. Hummingbirds do not live year-round in either the Sonoran or the Monte Desert because of scarce food during most of the hot, dry summer.

"Bats also pollinate," says Simpson, "but only in the Sonoran Desert, where they visit the saguaro cactus flowers for the nectar, rich in amino acid. However, the most important pollinators at both my field sites are solitary bees."

In both places some bees restrict their visits to only a few species of plants; other insects forage on a wide variety of flowers. Most of

Top, Cassia aphylla *from the Monte Desert. Center,* Fouquieria splendens, *or ocotillo; bottom,* Olneya tesota, *or ironwood; both of the Sonoran Desert.*

Above, Prosopis torquata, *or screw bean, of the Monte Desert. Botanist Beryl Simpson,* right, *crosses flowers of* Larrea tridentata, *or creosote bush, in the Sonoran Desert.*

these plants provide nectar and pollen as "rewards," or attractors, assuring that they will be pollinated by insects. One Sonoran genus, *Krameria,* offers the females of the one bee genus that visits it a unique reward—oils secreted by the fleshy glands in its attractive pink or purple flowers. The oils with pollen are used by bees as the food supply in the subterranean nest cells where the bee larvae develop. In Argentina, a similar bee-flower interaction is found in the genus *Tricomaria.*

Simpson's study, and many of the other investigations of particular animal groups or plant-animal interactions, show that although two areas look very much alike in vegetation and landscape, there are basic biological variations in the community patterns. The existence of such differences—subtle and difficult to pinpoint though they may be—must nevertheless be taken into account when ecologists attempt to manage or model similar desert areas in different parts of the world.

The Magnificent Foragers

VI.   Hard Stuff & Fiery

Deep down, the earth bubbles and boils, now and then spewing
forth that internal, molten heat. The mountain trembles, spumes of
smoke rise and glowing lava rushes out, burning, effacing, and, at
length, cooling down darkly.

Fountains of Fire

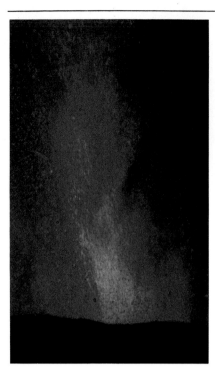

Lava sprays straight up, above, lighting the night sky on the Icelandic island of Heimaey during the eruption of Eldfell in 1973. Opposite, vivid rivers of lava snake away from Halemaumau, an active lava lake at the summit of Kilauea in Hawaii.

Fernandina, Galapagos Islands

The Galapagos Islands straddle the Equator some 600 miles west of Ecuador. To some observers, the fame of their unusual wildlife has overshadowed the fact that they are a major volcanic field which, incidentally, has been particularly restless of late.

In 1968 one of the spectacular volcanic events of this century shook the Galapagos island of Fernandina when the floor of the crater of the volcanic island erupted and collapsed. Calderas—large craters or depressions in volcanoes—are often filled with water, as in the case of Oregon's Crater Lake. But when a caldera collapses, catastrophic consequences often follow. The largely submarine collapse at the Greek island of Santorin is believed to have produced sea waves that devastated the Minoan civilization in about 1470 B.C.; similar waves from the famous 1883 eruption and collapse of Krakatoa, west of Java, killed 36,000 people. The collapse at Fernandina was much smaller than Santorin or Krakatoa, but it is second in this century only to the 1912 collapse of Katmai in Alaska.

Smithsonian volcanologist Thomas E. Simkin, who specializes in studies of volcanoes that erupt in mid-ocean, has been able to reconstruct the catastrophic occurrence at Fernandina. "It started on June 11, 1968, when a huge explosive eruption blanketed the central part of the volcano with over 50 feet of volcanic ash and debris and sent up a dark cloud that later dropped fine ash on the deck of a ship 200 miles away," he recalls. "Fernandina is uninhabited by humans, but both plant and animal populations suffered."

The eruption removed the support from the central part of the caldera and triggered the collapse. It soon began to subside, dropping 1,150 feet over the course of the next nine days. The floor was a thick block, and it jolted down nearly 12 feet at a time, each major jolt causing earthquakes that were registered all over the world.

"We at the Smithsonian first learned of the Fernandina collapse because of the mighty explosion that accompanied it," Dr. Simkin continues. "It was heard hundreds of miles away on other islands of the Galapagos chain where listeners said it was so loud that it sounded like a large dynamite blast a few blocks away. Sound waves from the explosion were recorded by infrasonic sensors from Bolivia to Alaska."

A party of scientists from Darwin Station, a laboratory on one of the nearby islands, visited Fernandina a few days after the shock and climbed to the crater rim. But so much dust rolled up from inside the crater that they could see little. On July 4, Simkin, his wife, and two other scientists—a geologist and a biologist—reached the islands by U.S. Air Force plane.

"The plane landed on one of the central islands and we had to get to Fernandina by boat," says Simkin. "After a two-day sail we arrived, set up base camp, and then hiked up to the rim. It was three weeks after the collapse started, but there were still minor earthquakes in progress as the caldera adjusted itself to its new topography. Huge rock avalanches were continually roaring down the oversteepened sides of the caldera. It wasn't safe to go down inside, so we contented ourselves with tramping around the rim, doing a lot of crude surveying and taking samples of ash.

The Magnificent Foragers

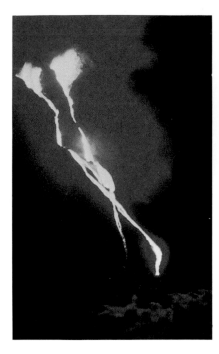

Above, lava flows some 2,000 feet from the vents at Fernandina in the Galapagos. Right, Smithsonian volcanologist Tom Simkin stands on the cracked and blistered floor of a playa, dried bed of a lake formed in Fernandina's recently collapsed caldera. Behind him, still steaming, is the vent ridge of the major eruption which triggered the collapse.

"Caldera collapses are so rare that they have seldom been scientifically documented. The Galapagos event has proved a remarkable opportunity to learn something about processes that have shaped one of the world's more interesting landscape features. We've been trying to fit together the various pieces of the puzzle by geological detective work—trying to understand the nature of the main eruption and the ones that followed in 1972, 1973, and 1977."

In four subsequent surveys, Simkin descended into the collapsed caldera. "We think that the successive volcanic eruptions over millions of years may have played a role in shaping some of the weirder characteristics of Galapagos wildlife," he says. "One of the things we're studying is the effect of the recent eruptions on the life in the area, which we notice is already making a comeback. The new lake at the bottom of the caldera is full of algae; ducks and other birds have returned; and large iguanas can be seen down at the edge of the lake. Life does have this irrepressible way of recovering."

*   *   *

On his return from the 1968 Galapagos shocks, Simkin stopped in Costa Rica to investigate the eruption of Arenal, a volcano long thought to be extinct. A colleague of Simkin, the Museum's William G. Melson, arrived two or three days later. Without warning, Arenal had exploded and buried a town at its base in an avalanche of ash and lava, killing 80 persons. "The Arenal eruption was an example of a volcano that everyone thought was dead, but wasn't," Dr. Melson explains. "We were able to tie Arenal's last eruption to the time of the Spanish arrival in Costa Rica through pottery we found buried in

Still related to big bangs, brass gun shells, above, have been adapted to house measuring instruments used by volcanologists. Thought to be extinct, Costa Rica's Arenal volcano, below, erupted without warning in 1968. Lava flows continued to advance in 1971, when this photo was taken.

Hard Stuff & Fiery

layers of ash and which archeologists later identified for us. There are many other volcanoes that explode at intervals of 300-500 years, and aren't generally recognized as being potentially dangerous. For example, we've only had 100 years to observe many of the volcanoes in the western U.S. that are thought to be extinct."

According to Melson, Arenal and Fernandina represent two distinct volcano types. The former is typical of volcanoes of the continental margins which are usually rather explosive and produce more ash than lava. The ocean basin volcanoes, like Fernandina, are less explosive and dangerous, yet produce more lava than ash.

Another major ocean basin volcano, Iceland's Heimaey, erupted on January 23, 1973. Learning of it, Simkin caught the first plane that evening, reaching the site on the second morning of the eruption. On the third day he was able to get to the town of Heimaey with some Icelandic scientists and find a vantage point only 700 yards from the lava fountain. Before ashfall forced him to evacuate the area, Simkin photographed and collected samples of ash that had fallen on a small area over a measured length of time. He bagged these and brought them back to the Smithsonian where he recorded the chemical characteristics of the ash—a record important to an understanding of the evolution of this ocean basin volcano.

Volcanic activity at Heimaey was caused by the sea floor splitting and widening along the Mid-Atlantic Ridge, a major mountain system in the Atlantic Ocean. As the rifting occurred—at the rate of one to two inches a year—molten volcanic rock called basalt welled up from the earth's interior to fill the cracks, giving rise to submarine volcanic eruptions.

Melson has served as one of the scientists-in-charge aboard the

Sulfur crystals, above, line the jagged edges of a fumarole, or gas vent. A spatter vent, below, percolates with lava and gases. Opposite, a geologist plumbs the depth of a Hawaiian lava river through a "skylight" in the roof of the lava tube.

Glomar Challenger, the National Science Foundation-financed ship that has been drilling along the crest of the Mid-Atlantic Ridge to investigate the composition of the basalt forming there. In his log of July 4, 1974, aboard ship, he wrote: "An intense morning—looking at our longest core so far—22 feet of gloriously interesting material."

Melson was especially pleased that day because the ship's drill had reached 563 meters (1,847 feet) on its way to a historic, still unsurpassed, 583.5-meter (1,910-foot) penetration of the ocean floor. The drill pierced the sediment layers that overlie the ridge's volcanic basalt and continued on down into the ocean's basement rock.

Glomar Challenger was operating in 6,000 feet of water, and the technological difficulties involved in holding a bobbing, rocking ship steady while it guided a drill-string two miles down to the ocean floor were staggering. Making it possible were computer-directed propellers that could move the ship sideways as well as backward and forward to keep it in place, and a funnel-like sonar-homing cone that was lowered onto the drill hole site on the ocean floor. When one of the diamond drill bits became dull, the drill-string was pulled back up to have the bit replaced; then the whole string was lowered again, and—with the help of sonar—guided through the cone and back down into the hole.

Overall, the drilling went smoothly, but occasionally there were problems. At one hole the drill-string casing broke just below the re-entry cone. At another hole the bit became irretrievably stuck at a depth of 312 meters and had to be blasted free.

Drilling went on around the clock during the cruise. Below decks was a fully equipped laboratory in which Melson and other scientists

Thick smoke and ash pillar up on the fourth day of an eruption at Heimaey, Iceland. Helgafell, the peak to the right, was formed during an eruption 5,000 years ago.

The Magnificent Foragers

Fountaining plumes of lava starkly silhouette charred tree skeletons at Kilauea in Hawaii during the eruption of 1969.

worked through the nights, describing and interpreting the new core material that was drawn up every hour and a half. These samples of igneous and sedimentary rocks are now undergoing further laboratory analysis at the Smithsonian and other major research centers.

"We're trying to figure out what's happening in the deep spreading areas of the ocean floor," says Melson. "We know that the crust is coming apart in these zones and that volcanoes are filling the center of the zones with new crustal material. By looking at the chemical makeup of the volcanic rock cores we're bringing up, we can see at what rate this new crust forms, and gain some understanding of seafloor spreading dynamics. A series of processes are forming new ocean floor, causing the movement of large blocks of the earth's surface, and controlling the drift of continents."

Hard Stuff & Fiery

The Volcano That Didn't

Guadeloupe

In August 1976 La Soufrière, a volcano on the island of Guadeloupe in the French West Indies, appeared ready to erupt. Dense clouds of ash were spewing from the volcano, and hundreds of earthquakes were recorded each day. Fearing that a truly devastating crisis was about to occur, French authorities ordered the immediate evacuation of all 72,000 persons living on the flanks of the volcano. A French scientific team was on the scene, and at their invitation Smithsonian volcanologist Richard S. Fiske flew in to assist in monitoring the volcano's activity.

At Dr. Fiske's suggestion, tilt monitoring stations were installed on the flanks of the volcano. (If tilt occurs, it is a warning that the volcano is being inflated by rising magma and is in danger of erupting violently.) Rods 40 to 50 meters apart, forming a square, triangle, or line, were set into the slopes and optically surveyed to detect slight changes in ground tilt. The scientists, however, detected no significant swelling of the volcano that would have indicated new magma was inflating the volcanic edifice. Without positive indications that the volcano was building toward a large and violent eruption, the government began to allow the evacuees to return to their homes. In early 1977 the volcano's tempo of activity began to dwindle, and by spring things had returned to normal.

Crews hammer solid aluminum rods into the ground, above, to measure volcano deformation at La Soufrière in Guadeloupe. Below, the old fortress which was converted to bunkers for the French teams that came to study the anticipated eruption. Outfitted with special air conditioning systems and generators, the fortress was equipped to function despite volcanic activity.

After an ominous build-up, the fissures at Guadeloupe merely belched forth steam and wet ash. Below, a bedraggled technician, grimy with wet ash, hauls his equipment into the compound at the fortress. Though anticlimactic, the non-eruption was no less significant to scientists, helping to create better profiles of volcanic activities and warning signals.

Night of the Bolide

This photograph has been enhanced by an artist to show the fall of a meteor over the Prairie Net tracking station run by the Smithsonian Astrophysical Observatory, with help from NASA. Using photography, tracking stations throughout the Midwest could triangulate to pinpoint impact locations of meteorites. The Lost City Meteor was successfully retrieved using this method.

Pueblito de Allende, Mexico

It was just after midnight on February 8, 1969, when the slumbering inhabitants of the northern Mexican village of Pueblito de Allende were routed from their beds in terror by the brilliant light and booming detonations of a fireball passing overhead.

When news arrived at the Museum that a shower of meteorites had landed in northern Mexico, Brian H. Mason and his fellow curator of meteorites, Roy S. Clarke, Jr., flew to El Paso, Texas, where they rented a car and drove across the border toward Allende. "We arrived to find the people highly excited. Fortunately, no one had been injured, although one large fragment had come down only a few feet away from a village home," Dr. Clarke recalls.

By interviewing residents and pinpointing each find on a map, Mason and Clarke were able to reconstruct this violent episode. The meteor had come in low from the south-southwest, growing into a brilliant fireball and exploding into incandescent fragments that hailed down as individual meteorites. The largest fragments traveled the farthest north and shattered as they hit, making small holes in the ground. The smaller pieces—some of them only tiny grains—fell far to the south. The whole swarm of fragments cascaded within an elliptical impact field 30 miles long.

Although thousands of pieces had fallen, they weren't easy to find. "You had to search at least an acre of ground before you ran across a single fragment," says Clarke. "We talked to the local schoolmaster, and he adjourned classes to let us take his sixth-grade pupils out into the country to join the search. It was a good thing we had them along because even though I walked for miles I never turned up a thing. Dr. Mason was luckier, he found one the size of a football."

Clarke and Mason brought back more than 300 pounds of black, fusion-crushed, stony meteorite fragments, most of them purchased from local residents. Samples were immediately packaged and mailed to other meteorite study centers around the world for mineralogical, chemical, and physical examination.

Allende material is now being called a "Rosetta Stone of the Solar System," because it has provided scientists a key to deciphering secrets of planetary formation. Tests have dated material found in the meteorite as 10 million years older than the 4.5 billion-year age of the earth and the moon. This makes it the oldest mineral specimen ever found, and scientists theorize that part of it is made up of interstellar dust grains that predate the solar system.

One odd footnote to the event: of the 10 largest meteorites ever discovered on earth, three fell in northern Mexico; two of them, the Morito and Chupaderos iron meteorites, were found in southern Chihuahua State within 50 miles of the area where the Allende fragments came down. A Mexican scientist once suggested that the soil of his country seemed to have the peculiar property of attracting meteoritic material. Actually, according to Mason and Clarke, the explanation is much simpler: the soil of northern Mexico has long been geologically stable, and large meteorites that fell in ancient times have remained undisturbed until discovered in recent years.

Mason and Clarke urge that anyone seeing a fireball should try to

Brian Mason, above, examines a chunk of Allende meteorite found by local people in San Juan, Mexico.

remember what time it appeared, its direction of travel, how long the light phenomenon lasted, what the color of the light was, and what kind of sound the fireball made. A person who believes that an object he has found is a meteorite—whether or not he witnessed the fall—should bring or send it to the Smithsonian for a free examination.

"Usually these objects turn out to be slag or some other heavy artificial or natural matter," says Clarke. "One exception was our big Clovis, New Mexico, meteorite. A man wrote the Museum and said he had dug up something on his farm that had to be a meteorite because there weren't any rocks where he lived. We asked him to send us a small piece. It turned out the man was right—he had unearthed a 600-pound stony meteorite."

The Allende and much of the other meteoritic material in the Smithsonian's collection is either scarce or unavailable anywhere else in the world, and it is highly sought after by researchers. Clarke scans requests for loans every week, granting them if the scientist is from a reputable institution and if the experiments he wants to perform with the material seem important. If Clarke believes that the collection can be enriched by doing so, he can recommend that material be exchanged with an institution or individual. The Soviet Government recently was granted a sizeable sample of one of the Allende fragments to use for research, agreeing in return to send the Museum a large piece of the Sikhote-Alin iron meteorite that fell in Siberia in 1947.

127   *Hard Stuff & Fiery*

Impact Craters

Lonar Crater, India

The Smithsonian's involvement with meteorites and mineral sciences goes back to its earliest days. James Smithson, the Institution's founder, published numerous mineralogical papers and amassed a collection of more than 5,000 mineral specimens that included a number of meteorites. Sadly, the whole collection was lost when a fire burned out part of the original Smithsonian building in 1865.

A dramatic increase in the size of the collections of rocks, minerals, ores, and meteorites came after the founding of the Smithsonian's U.S. National Museum in 1879. Material was deposited by the U.S. Government land surveys, private collectors, and by the Smithsonian staff. By the turn of the century, it was on its way to becoming one of the world's pre-eminent collections.

But two Mineral Sciences Department researchers, Kurt Fredriksson and Robert Fudali, concentrate not on meteorites per se but rather on the impacts giant meteorites make when they hit the earth. Dr. Fudali has completed a number of expeditions to Africa to study ancient impact craters, while Dr. Fredriksson's work has focused on Lonar Crater in India, a huge, rimmed, circular depression with a shallow lake in the central portion that appears to have been formed by a volcanic eruption. The only large meteorite crater in volcanic rock on earth, it may be directly compared to similar formations on the moon. Fredriksson's data indicates that Lonar Crater was produced by a gigantic meteorite that hit the earth probably sometime between 30 and 50 thousand years ago.

A saline lake fills India's Lonar Crater, situated in western Maharastra state and measuring 6,000 feet, rim to rim. Opposite top, the Smithsonian's Robert Fudali, standing, and a University of Capetown scientist conduct a topographic and gravity survey on the rim of Roter Kamm Crater. The crater, almost buried in red sand, lies in the restricted diamond area, about 60 miles north of the mouth of the Orange River, Namibia.

This stone building at Gros Brukkaros Crater in Namibia is a relic of science, the remains of a solar observatory built and operated by the Smithsonian from 1929 until 1932. Once thought to be an impact crater, Gros Brukkaros is now known to have been caused by explosive volcanism.

The site proves valuable because it provides the opportunity for a detailed study of debris ejected from a relatively recent meteorite impact in volcanic rock. According to Fredriksson, it has not been possible to obtain such detailed field data from the moon's lunar craters. "On the moon's surface," he explains, "we're left with a record of a period of intense meteoritic activity that took place during the formation of the solar system 4 to 4.5 billion years ago. During this time, as soon as one crater was formed, additional meteorites would land on and around it, mixing up the debris and creating a very complex and confusing history." This continuing meteoritic bombardment, which now takes place at a much lower rate, formed the moon's characteristic breccia rocks, which consist of mixed fragments embedded in a fine-grained matrix.

"In order to appreciate the magnitude of the forces involved," says Fredriksson, "consider the fact that a kilometer-sized meteoritic body traveling at 20 kilometers per second packs potential energy perhaps 10 times greater than the total amount of energy released in the eruption of volcanoes during each year on earth."

For 500 million years, such impacts may have been the dominant geological process on all the inner planets of the solar system, including the earth. We don't see the traces of these ancient impacts on earth because they have been obliterated by geological processes.

Hard Stuff & Fiery

Hard Stuff and Beautiful

Washington, D.C.

Specimens of amethyst, diamond, and hundreds of other kinds of precious and semi-precious natural crystals and gems collected from all over the world are stored in special Smithsonian rooms under tight security. The acquisitions in 1926 of the Canfield and Roebling collections, the two finest private mineral collections in the United States at that time, thrust the Smithsonian into international prominence as a mineral depository. Today, under the curatorial control of Paul E. Desautels and John S. White, Jr., the collection numbers more than 100,000 individual specimens and continues to grow in size, quality, and scientific importance.

"This isn't the biggest collection you can name," cautions Desautels. "I tell people that if they want to amass the world's biggest mineral collection, just hire a bulldozer and go scrape up 49 tons of rock. Size isn't a criterion. Quality is. And by any valid measurement, we think of this as the greatest mineral and gem collection in the world. Its many species and varieties, its worldwide representation, its current research value, its intrinsic value—put it all together and there's nothing to match it."

The collection's global reputation helps attract more materials. "People are turned on by the idea of preserving fine materials in the National Museum's archives," says Desautels. "Not long ago an amateur rockhound poking around in an abandoned feldspar mine in Maine tapped a rock wall with his geologist's hammer and broke through into the largest and highest quality pocket of tourmaline crystals ever found in America. The finder notified the Museum imme-

These large, natural gem crystals and cut gems, selected at random from the extensive gem and mineral collections, stand between 6 and 10 inches high.

The Magnificent Foragers

The world-famous W. R. Warner crystal ball, above left. Cut from a block of Burmese quartz estimated at 1,000 pounds, this extremely valuable, flawless, colorless sphere has a diameter of 12 ⅞ inches and weighs 106 ¾ pounds. Above right, a stream-tumbled nugget of gold penetrates the quartz in which it was originally formed. Weighing several ounces, this specimen came from an old gold occurrence in Spotsylvania County, Virginia. A polished sample of jasper—a fine-grained quartz— from Oregon, bottom right, resembles an unreal landscape. Many jasper patterns of great beauty and complexity exist in nature and are well represented in the national collection.

diately and selected for its collection some of the biggest and best of the cylindrical green and pink crystals found in the mine."

Was it important to add these to the tourmalines already in the collection? "Yes," responds Daniel Appleman, the Smithsonian scientist who specializes in studying the crystallographic structure of minerals. "If you try to draw conclusions about the structure or composition of a mineral species on the basis of one sample, you're often in trouble. Variances frequently occur. Occasionally two crystals that look the same and have been lumped together as such are actually two distinct species. A big, systematic collection like this helps make such distinctions.

"Before you can be certain that you have characterized a true mineral species, you've got to analyze material from a wide variety of environments on the microprobe and find a relative chemical uniformity. We've had the equipment for microprobe and x-ray diffraction analysis only during the past decade. The traditional wet-chemical analytical methods were often inaccurate, and a lot of material has been incorrectly described. Our collection is a reference tool,

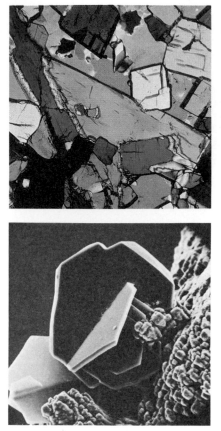

Scientists of the Mineral Sciences Department have long been mineralogically and gemologically interested in jade. This sample of Burmese jadeite jade, top, has been ground to a fraction of a millimeter thickness, magnified 100x, and photographed through a petrographic microscope. Bottom, a scanning electron photomicrograph of desautelsite, a new mineral species from Cedar Hill Quarry, Lancaster County, Pennsylvania. This new mineral was discovered and characterized in the Smithsonian's Mineral Sciences lab by Pete J. Dunn.

more or less an encyclopedia of mineralogy, and it is important for it to be accurate. That's why we need so many of the very best crystals. The best specimens are always the most useful to analyze."

According to Desautels, his problems in acquiring mineral and gem material are quite different than those of the other curators in the Museum. "Biological specimens often arrive in large quantity through field research," he observes. "Unfortunately, fine minerals and gems can't be plucked off trees like some botanical specimens."

Other Government agencies such as the U.S. Geological Survey and NASA deposit material with the Museum according to law, but much of the very limited quantity of choice material needed for study isn't in their hands. The Smithsonian finds itself in direct competition with the three million rockhounds in this country—a group that ranges all the way from agate collectors who only spend a few cents on their hobby to millionaire connoisseurs.

The good will of the Museum's network of contacts assures some choice acquisitions—as in the case of the tourmalines—but much of the showy, top-quality specimen material that the public enjoys seeing on display is very difficult to come by, especially the gemstones. "It's just like being in the art market, and sometimes the transaction boils down simply to money," Desautels says. "This leaves two routes open to us: taking a series of small, less significant specimens and exchanging them for the kind of item we need; or seeking an outright donation. To accomplish this, we show the owner that the donation is needed and that it is feasible for him to give it to us.

"We've been very fortunate in getting such donations. The month in which something significant doesn't come in is now a rarity. We keep an eye out for this good material by attending national and international gem and mineral shows and conferences. This keeps us in touch with what's developing in mining, and if anything new appears or a major collection goes on the market, we're always aware of it very quickly. All of the major dealers are our friends, and when one of them uncovers something good, he will likely inform us first."

The most notable gemstone ever to be donated to the Museum was the Hope Diamond, reverently placed in its illuminated glass wall case on November 10, 1958, a gift of New York jeweler Harry Winston. It was the first major gem in the Museum's collection, and according to Desautels' colleague, George Switzer, the very special aura it radiated helped draw other important gemstones. "Before 1958 we didn't have the very best display facilities," Dr. Switzer recalls. "But the same year we got the Hope, we opened up a handsome walnut paneled exhibit hall. When the new hall opened it was the first time gems had been displayed in such an attractive setting. People like that. Then the Hope came. It gave us the major gem we needed to put us on the map. That's when more donations started to come in. The whole thing began to snowball, and before long we had the Portuguese Diamond, the Rosser Reeves Ruby, and the Star of Asia Sapphire. We were on our way."

Just recently a gift of the glamorous, champagne-colored, 67.89-carat Victoria-Transvaal Diamond has continued the tradition. Desautels anticipates that this new impetus will encourage more significant donations for several years.

The Moon and Tektites

National Museum of Natural History

Younger than meteorites but older than most rocks on earth, the rocks that the Apollo astronauts brought back from the moon provide a unique record of the first billion years of planetary history.

One day, the Smithsonian's Brian H. Mason, sifting through an Apollo soil sample shortly after its arrival from NASA's Lunar Receiving Lab, came upon a tiny metallic pellet. Dr. Mason turned it over to his colleague Edward Henderson, a metallurgist and meteorite authority. Dr. Henderson in turn placed it under a microscope and realized that he had something intriguing in hand—a minute meteorite, the first collected on the moon.

The pellet was named the Mini Moon because of its shape. "We concluded that it was created when a meteoritic fragment crashed into the lunar surface," says Mason. "The impact produced a liquid drop which solidified as it cooled in the lunar gravitational field and formed a spherical pellet. The pellet was then abraded and cratered by high-velocity lunar dust and other fragments until, curiously enough, it came to resemble man's image of the moon."

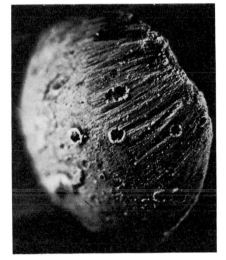

This small meteorite fragment, dubbed the Mini Moon, was discovered in the lunar material brought to earth by Apollo astronauts. Measuring only 4 millimeters in diameter and weighing 88 milligrams, it was originally part of a meteorite that crashlanded on the moon's surface. Subsequently, abrasions by lunar dust and high-velocity lunar particles produced minute craters.

The Mini Moon was only one of many discoveries made by Smithsonian investigators during their examination of the moon rocks and soils. On the whole, the laboratory tests on lunar material are the same as those applied to meteorites and terrestrial rocks. The tests begin when chemists take a small sample for bulk chemical analysis to establish its elemental constituents.

Thin polished sections of the sample are then produced with the aid of a diamond-bladed saw and buffing wheel. The sections are subsequently mounted and examined under both standard optical microscopes and an electron microprobe. These and other instruments identify the minerals in the lunar material and present a picture of how the chemical elements are distributed within these minerals. In all, the Smithsonian staff has identified more than 20 minerals in the moon material. They include at least one new mineral—an iron calcium silicate which occurs in small amounts in many of the moon's crystalline rocks.

The resulting analysis permitted the researchers to divide lunar rocks into two main types: volcanic rocks, lunar basalts chemically and mineralogically different from earth basalts; and microbreccias, glass and rock fragments melted and consolidated into a coherent mass by heat generated by meteoritic impacts.

X-ray diffraction equipment is also employed in the analysis, a laboratory operation supervised by the Smithsonian's crystallographer, Daniel Appleman. This examination provides a picture of the structure of each mineral within the sample, illustrating the geometric relationship of atoms in the minerals.

"The Smithsonian's principal interest is the analysis of the lunar material's mineralogical and chemical composition, and its structure and classification," says Mason. "When these facts are established we have the sample's basic pedigree. This is the same procedure that we apply to meteorites and terrestrial rocks."

When it comes to tektites, however, Mason and Henderson lack a critical piece of information: where did tektites come from? Tektites

Hard Stuff & Fiery

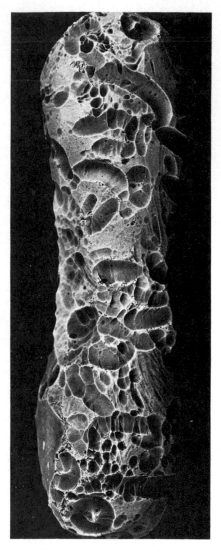

are black or green glassy droplets found in a number of streamlined shapes—spheres, teardrops, dumbbells, and flanged buttons. Scientists believe they were shaped while flying through the air in a molten state. But their source is still a mystery. Were they splashed into orbit by a huge meteoritic impact on earth, only to fall back into earth's atmosphere? Or are they pieces of the moon, flung to the earth when ancient lunar volcanoes exploded, or when mountain-sized meteorites blasted great craters?

As yet, no site has been found on the moon that matches tektite chemical composition, but much of the moon surface remains unexplored. It would certainly seem that a crater from an impact or explosion on earth large enough to provide tektites would still survive. Some scientists have suggested that such a crater may exist beneath the Antarctic ice cap.

Whatever their origin, we know that billions of these objects—as much as 10 to 100 million tons of material—probably were formed by a single event and were scattered in a vast S-shaped pattern extending from near Madagascar to Tasmania, northwest across Australia, north over Indonesia and northeast over Asia and the Philippines. An overwhelming percentage of this hail of tektites splashed into the ocean, where they are occasionally dredged up by fishermen. Probably the best place to search for well-preserved ones is on the flat interior plains of central and western Australia. During five field trips there, Mason and Henderson have acquired the best-documented collection of Australian tektites, more than 2,000 of them.

Carbon-14 datings of calcium carbonate taken from the beds where the tektites were found point to an age of 12,000 to 16,000 years, suggesting that they fell relatively recently. But analysis by other scientists, using potassium-argon dating methods, contradicted these figures, indicating instead that the tektites were shaped from molten material 700,000 years ago. "There just isn't any certainty about either their origin or their age," says Mason, shaking his head. "In spite of all the work we've done on tektites we still don't have any solution. It's a real scientific whodunit."

The elongated and deeply grooved tektite, above, was found in gravel on Billiton, an island between Sumatra and Borneo. Groundwater seepage caused markings. A morning's collection of tektites, right, from a site near Pine Dam, South Australia.

The Magnificent Foragers

A Meteoritic Glossary

Comet:
Collections of frozen gases (methane, ammonia) and "earthy" materials that travel in highly elongated orbits around the sun. Comets usually are invisible until they near the sun. Then their materials are activated and produce a coma of gases and dust that becomes a tail extending, in some cases, millions of miles.

Asteroid:
A relatively small planetary body orbiting the sun. Most asteroids are found in the asteroid belt between the orbits of Mars and Jupiter. The largest asteroid, Ceres, is only 770 kilometers in diameter.

Meteoroid:
Cometary or asteroidal debris in orbit about the sun. The meteoroid becomes a "meteor" as it enters our atmosphere and becomes a "meteorite" if it reaches the earth. The differences are both semantic and physical since all three terms designate different stages in the life of the same object.

Meteor:
When a meteoroid enters the earth's atmosphere, it generates light, either from the glowing of the body itself caused by friction, or from the luminescence of the glowing gases surrounding the body. Technically, this light itself is the "meteor."

Meteorite:
Meteoritical material that has fallen on earth, ranging in size and weight from tiny stones to mammoth objects weighing 60 tons or more. The microscopic particles of cosmic dust falling on earth are called "micrometeorites."

Annual Meteor Shower:
Annual meteor shower occurs when the earth's orbit passes through a stream of cometary debris also in orbit around the sun.

Meteorite Shower:
A meteorite shower or swarm occurs when a single meteor breaks into several parts in its journey through the atmosphere.

Fireball:
A meteor of extreme brightness.

Bolide:
A meteor that explodes in its descent to earth.

Shooting Star:
A popular name for meteors.

VII.

Movable Seas & Oceans

Subtle signs of change can be read in the tell-tale shells of perished creatures of the deep. Lesson: given enough time and the restlessness of a tectonic earth, seas and oceans migrate between continents.

Canyons of the Deep

Mediterranean Sea

For many years oceanographers bypassed the Mediterranean and concentrated instead on studies of the world's larger seas. Only relatively recently have they begun to map out the Mediterranean's undersea landscape and reconstruct the processes that shaped it. One of the most mysterious features of its deeply hidden contours are numerous V-shaped gorges. These originate off the Mediterranean's coasts and wind their way seaward across its steep continental slope to the abyssal plains beyond. Similar undersea canyons occur off the U.S. Atlantic Coast and in other oceans.

To learn how these canyons are structured and how they serve to trap and channel sediments from rivers, shorelines, and the adjacent continental shelf, Smithsonian oceanographer Daniel J. Stanley has descended 10,000 feet in a research submersible, peering out of its portholes to study steep-sloping canyon walls. On other occasions, using an undersea TV camera and lights lowered off the stern of an oceanographic vessel, he has scanned the bottom of canyons on a video monitor.

"In both circumstances," he explains, "I could see clouds of suspended sediment streaming by, fragments of clay and sand borne by surprisingly strong bottom currents that were distributing sediments on the canyon floor as well as eroding it. Wrappers, toilet paper, beer cans, and wine bottles, some probably dumped overboard from passing ships, are evidence that man is increasingly adding his own sediments to the sea bottom. Even so, I could also see a surprising variety of life: lobsters, crabs, fish, and worms; animals that burrow deeply into and rework the bottom sediments, causing failure of the canyon walls and the eventual slumping of massive amounts of sediment down the continental slope toward the abyssal depths."

Theories on the origins of these canyons have been long and hotly debated by marine geologists. One recent view is that they were eroded by sandy underwater flows called turbidity currents. But Dr. Stanley's observations suggest that these processes empty the canyons but do not carve them. More important to their creation, and clearly underestimated, is the role of sea-floor spreading, which initially pro-

Smithsonian oceanographer Daniel J. Stanley, entering the two-man submersible Nekton Gamma, *above. Right, the submersible returns to the ship after a dive in the Hudson Canyon off the U.S. East Coast.*

Top, a bottom camera system being lowered off a research vessel in the Strait of Gibraltar. Below, the shell-covered sea floor at a depth of about 900 feet near the head of Hudson Canyon. This view, seen through the submersible port, reveals crabs, starfish, and sand-filled clam shells.

duced major tension faults (cracks) on the sea floor, some of which developed into valleys on steep continental slopes. The canyons were deepened during the ice ages, when vast quantities of water were withdrawn from the ocean and locked up in polar glaciers. At those times, sea level was lowered considerably, and large areas of the continental shelf became exposed.

As a result, our continental river systems extended much farther seaward than they do today, and the paths they often followed and carved out were the fault traces. Another factor played an important part in a canyon's complex structural and erosional history. Continental margins have continued to settle during long periods of time and this has lowered river-cut valleys considerably, thus accounting for the presence of submarine canyons to depths of two miles or more below sea level.

Stanley's Mediterranean research began in the 1950s when, as a doctoral student, he began a study of the geological history of the French Alps, a project prompted by his curiosity about the origin of this spectacular mountain range, as well as by his interest in mountain climbing.

The Alps are made up of thick layers of ancient marine sediments now uplifted 10,000 feet or more above sea level by geologically recent crustal movement. "As I studied sedimentary rocks on my climbs in the mountains," says Stanley, "I began to realize that at one time they accumulated under conditions likely to be found in the Mediterranean and similar geologically mobile oceans. I knew that if I were to understand Alpine geology, I would have to understand the factors that influence the way sediments accumulate in the Mediterranean today."

Very little sea-floor research had been done at that time in the Mediterranean. Existing knowledge stemmed largely from the efforts of the U.S. Navy, which had taken bottom soundings for anti-submarine warfare purposes during and after World War II. These surveys and the studies of the French oceanographer Jacques Bourcart provided the first reliable relief maps showing the highly complex character of the Mediterranean's shelves, steep continental slopes, submarine canyons, and basin floors.

By the end of the 1960s, deep-sea TV cameras, sonic profiling, and research submersibles became available as research tools. As a result of the famous *Glomar Challenger* deep-sea drilling project, it also became possible to collect sub-seabottom soundings of the sea floor and recover deep sediment cores.

"The more I study the Mediterranean," Stanley goes on, "the more I realize that it is a fantastic natural sedimentation laboratory. It is a very compact sea, but highly complex in its configuration. Its exchanges of water with the Atlantic and Black Sea are so limited that it functions virtually as a closed physical-chemical-biological system. This helps make it possible to interpret accurately where sediments have come from and how they have been moved."

Since many of the processes that govern the movement of sediments also determine the dispersal of pollutants, it follows that knowing what happens to sediments when they are introduced into the system makes it possible to predict where some of the pollutants will go and, importantly, what their consequences will be.

The Magnificent Foragers

Tethys Ocean

Mediterranean Sea

The largest ocean systems in the world today run north and south, distributing the cold waters of the polar regions throughout their vast abyssal depths. But this was not always so, according to the Smithsonian's Richard H. Benson, who has discovered fossil evidence that major changes have taken place in these world ocean systems over the past 100 million years. "There was another ocean system that no longer exists," he says. "It was as large as the present Atlantic but had a dominantly east-west circulation pattern. It was a source for warm abyssal waters. The last days of its distribution occurred within the history of earliest man, and were probably seen by him."

The modern Mediterranean Sea fills what is left of the western part of the many basins of this vanished body of water. Called Tethys Ocean, after Greek mythology, its shores stretched from the area of the China Sea, followed the present lines of the Himalayas and Caucasus Mountains through the Mediterranean region, and, running along the same latitudes, crossed between the two American continents to empty again into the Pacific. As the Atlantic Ocean was formed, Tethys disappeared, destroyed by giant movements in the earth's crust.

Dr. Benson discovered the former existence of this ocean by studying a tiny, bottom-dwelling crustacean called an ostracode. The microscopic skeletal remains of this creature are found in the muds and rocks of both present and former seas and oceans all over the

The northwest coast of Sicily, where the deep-sea floor of 5 million years ago has been thrust upward to form rugged mountains. Below, Richard Benson collecting microfossil samples from upturned strata. The fossils are an indicator of oceanic history.

Movable Seas & Oceans

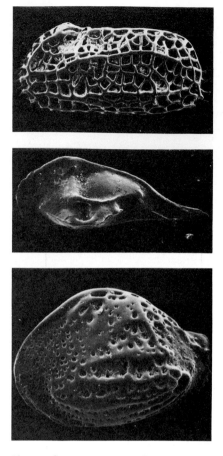

Shown above in scanning electron microscope images, the intricate architecture of ostracodes—fossil crustaceans—is associated with the history of an ancient global ocean named Tethys. Unlike present oceans, which circulate north-south, its waters circulated east-west. The maps, opposite, illustrate the patterns of warm water production during the east-west circulation period 75 million years ago, top, and the subsequent shift to north-south cold water currents 30-35 million years ago. Color indicates temperature: orange warm, blue cool.

world. They are now being recovered in ocean-floor drill cores taken by the *Glomar Challenger,* the research ship of the International Program for Ocean Drilling.

More than 30,000 different kinds of ostracodes have lived over the past half billion years. Some of the structural changes undergone by many of them reflect variations in temperature, oxygen content, wave action, and depth of past oceans and seas in which they lived.

"With the aid of the scanning electron microscope and the computer, it is possible to examine the design of ostracode carapaces (shell-like exoskeletons) and to fit the pieces of their ancestry back together again," states Benson. "We can see that, if the ostracodes are blind, they may have lived in the depths below which light can penetrate; and if their carapaces are thin and designed to compensate for the lack of calcium carbonate, then the temperatures of the water surrounding them were probably cold. With a number of such comparisons, and a knowledge of modern oceanography, we can arrive at a fairly accurate estimate of whether the ostracodes we find in cores once lived in warm, shallow seas or in deep, cold oceans."

Benson first discovered oceanic ostracodes in the mountains of Sicily, in rocks nearly 5 million years old. The discovery of these ostracodes, at a site more than a thousand miles from their nearest living relatives in the Atlantic, led to his search for more evidence of the lost Tethys Ocean. Though geologists had known about Tethys for almost a hundred years, they had imagined it as a shallow sea with a few deep pockets in the area of the Alps—not as a deep ocean system connected to all of the other ocean systems. The fact that Tethys was not simply a flooded continental region (a sea by scientific definition) but that its waters actually filled the crustal void between continents is of great tectonic significance.

Tectonics is the science of the earth's crustal movements. Some parts of the crust move up and down, some sideways past each other. But the greatest movements are those of large sections called "plates" that move whole continents and parts of the ocean floor. In 1965, plate tectonic theory, which explains how some oceans form as continents drift apart and others are destroyed as continents drift together, was not the generally accepted scientific principle that it is today. Benson's suggestion that the western Mediterranean (thought by many to have been land 4 million years ago), Sicily, and indeed, much of Italy, lay at the bottom of a deep ocean only a brief geologic time ago was looked upon skeptically. But his claim was upheld when cores obtained by the *Glomar Challenger* in the Mediterranean, and samples of rock taken from outcrops in Italy and from sites as far north as Moravia in Czechoslovakia, all proved to contain oceanic ostracodes.

The fossil record traced by Benson shows that Tethys died a slow death under the "subduction pressure" of plate tectonic movements. Between Africa and Europe colossal forces pulled the southern plate under that to the north and pushed and shoved up the high, mountainous borderlands of the Mediterranean, shrinking Tethys in size and cutting off part of it by forming a warm inland sea. At its historical climax 6 million years ago, Tethys was almost, if not completely, cut off from the Atlantic, and its waters lowered and evaporated, leaving behind deserts, shallow lagoons, and tremendous evaporite deposits of

The Magnificent Foragers

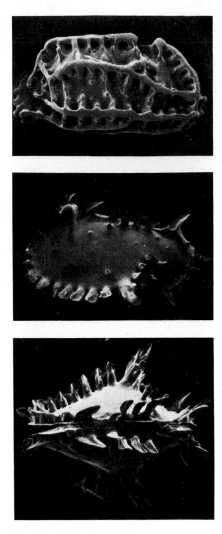

St. Peter's Basilica in Rome, opposite. Repair work on its dome led Dr. Benson to reassess the significance of architectural subtleties in the evolution of ostracode exoskeletons, more examples of which are shown above.

gypsum and salt nearly a mile thick in places. This phenomenon may have decreased world salinity nearly 10 percent and triggered the Ice Age. Then, about 4 million years ago, the dam formed by mountains near the Strait of Gibraltar ruptured, and the Atlantic reflooded the western Tethys basins, forming the oceanic stage of the present Mediterranean Sea. Ultimately the gap in the dam was nearly closed by further tectonic movements, isolating the Mediterranean and making it the warm "sea" it is today.

During all of the history of the destruction and extinction of Tethys, other massive plate movements and "sea-floor spreading"—changes that were all recorded in the evolutionary history of ostracodes—had brought into existence the present Atlantic Ocean. Beginning 135 million years ago, crustal plate movements split North America away from Europe, and South America away from Africa, to form a narrow seaway. Later, about 60 million years ago, the crustal plate with Antarctica began moving toward the South Pole. The complete north-south deep-water connection between the North Atlantic and the south polar seas was not completed, however, until about 38 million years ago when a ridge that was keeping Antarctic bottom waters out of the Atlantic collapsed or was overcome by the cooling southern waters. This event changed the stratification of water masses in the whole world ocean system and brought about a major evolutionary change in the ostracodes. With the lowering of a cold-water barrier and the gradual death of Tethys, the Atlantic evolved from a one-layer, warm-water ocean system into its present state as a two-layer, deep, cold ocean.

For the ostracode, the shift to the new ocean environment required a significant modification in its "architecture." The understanding of this change has its analytical counterpart in the understanding of the structure of buildings designed by humans, a discovery Benson made in an almost accidental way while admiring one of the glories of Mediterranean architecture.

"It is impossible to work around the Mediterranean and the remains of its ancient cities without becoming interested in how the science of architecture evolved," Benson admits. "While in Rome visiting a colleague in the Vatican, I had an opportunity to go inside the great dome of St. Peter's Basilica. Through time, the dome has cracked in places from stress. I noticed that an oak band had been fastened around the base of the dome's exterior to prevent more cracks from developing. This resolved the problem of 'thrust' created by the stones too weak to bind the bottom of the dome together. My interest in this principle led to the realization that the ostracodes were doing the same thing in order to maintain the structural integrity of their shells, and that their design reflected a response to stress created by environmental loads and pressures. Ostracodes are basically domed and have developed a sort of organic 'tension ring' around the margin of their calcareous shells to strengthen them. This tension ring serves the same purpose as the oak band in St. Peter's Basilica."

Other "construction" analogies can be made between ostracodes and different facets of engineering, according to Benson. Unlike mollusks and most other shelled invertebrates that gradually construct their shells over a lifetime, the ostracodes, which are arthropods, secrete a whole series of new shells, framing the design of each growth

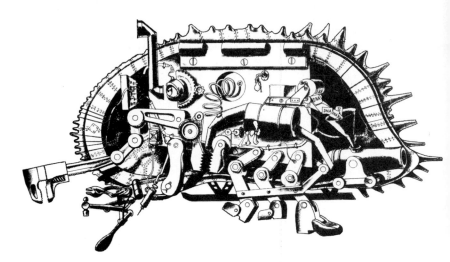

This whimsical cartoon represents an analytical and philosophical abstraction of the function of the internal organs of an ostracode.

stage by hydrostatic pressure, like pouring cement onto a balloon. This happens nine times during their life span, and the process increases their size each time by doubling their volume. Ostracodes expend a great deal of energy in forming and accumulating the tremendous amount of construction material they need for each new shell. Thus it is important for them to conserve the material in a very economical way, as would a good engineer, producing a design that is as efficient as possible. This building job also has a deadline. The animal is vulnerable while it is inflating, so construction of each new shell must be completed within a few hours. Fortunately for the paleontologist, many of the old shells that have fallen into the sediment have become part of the fossil record, a record which can lead to major discoveries.

"There is no doubt that they are sophisticated engineers," says Benson. "I found as I studied their mechanics that they had developed a limited number of geometric styles to achieve strength. Some of these design styles have been used and reused over millions of years. Among them we find well-known engineering structures such as the catenary, cantilever, corrugation, and static box-like frames. I found also that the patterns of ribs on their bodies change according to strength requirements. When the animals moved into the deep, cold Atlantic waters, for example, they evolved new shapes by perfecting old structural styles. I've now been able to encode the patterns underlying these structures for the computer, giving us a way to trace the ostracodes' evolution and discover the evidence for such major events as the formation and destruction of world ocean systems."

Recently, Benson has been searching for the ancient "Tethys-Atlantic connection" that existed after the modern Atlantic circulation system was formed and before the crisis event that ended Tethys. He has traced what he thinks is the most important part of this deep passage through southern Spain on the other side of the Sierra Nevada Mountains north of Gibraltar. There may have been another passage in the Pyrenees Mountains that separate France and Spain, or in the Atlas Mountains in northern Morocco. As chairman of the panel of paleontologists who advise on the drilling by the *Glomar Challenger*, he is hoping to see the other side of the picture as they drill into the margins of the North Atlantic in the next several years.

The Magnificent Foragers

Lampshells and Snowshoes

Hydra, Greece

Brachiopods still do not have a common name today, but ancient Greeks knew them and called them lampshells because they were shaped like their oil lamps. What makes them interesting to paleobiologists is that their line has survived through the entire 600-million-year period in which there has been shelly life on earth, leaving behind an enormously rich and complex fossil record.

In search of a large collection of fossil brachiopods for study, the Smithsonian's Richard Grant and his assistant Rex Doescher visited the Greek island of Hydra not long ago. Hiking across a high limestone ridge, they came to a spot marked on their maps. A steep incline let them look down 900 feet to a cove and the blue waters of the Aegean Sea. On the slope, sticking out of the bare-bedded limestone, they saw the edges of the rare silicified fossil shells they sought.

Two hundred and thirty million years ago during the Permian Period, when shallow seas covered most of the earth's continental masses, these fossil animals were part of a living reef. As millions of years passed, the reef was buried under layers of mud and ooze, preserving the shelly hard parts of the brachiopods and other dead reef animals in sediments that later turned to limestone. Then silicification took place, a phenomenon involving chemical interchanges with the surrounding sediment.

"We spent most of the day hard at work," Dr. Grant explained, "gouging slabs of fossil-bearing rock out of the slope, chipping the slabs down to manageable size with our hammers, placing them in piles, and measuring the slope's stratigraphy with a tape and hand level. The next day we returned with horses, loaded the rocks in burlap pack sacks, and hauled them to the village for shipment to Washington."

An old hand at collecting silicified brachiopods, Grant began collaborating with the Smithsonian's G. Arthur Cooper in the late 1950s. Fifteen years earlier, Dr. Cooper had revolutionized paleontological science by setting up a large-scale acid processing plant to prepare the rock he was collecting in the Glass Mountains of Texas—the world's richest deposit of silicified fossils.

Richard Grant holds the spiny brachiopod shown enlarged in the background photograph.

A field worker looks down over a bedding plane of Permian limestone on Hydra, above. Right, an outcropping of Permian rocks on the south side of Hydra. Brachiopods were collected nearby.

The Magnificent Foragers

The harbor and village of Hydra. On the dock in the foreground, brachiopods found in the surrounding mountains are being unloaded from mules and put onto waiting ships.

When Grant's shipment of Greek rock arrived at the Smithsonian, each piece was placed carefully into a tub of hydrochloric acid. The acid dissolved the limy rock into mud and sand which drain out through a screen at the bottom of the tub, leaving thousands of fossils for Grant to sort. Brachiopods at all stages of growth were tangled together in this mass, from tiny baby shells to large, thickened "granddaddies." Some were bizarrely beautiful, with whisker-like spines sprouting from their shells.

"Unless the specimen is etched out of the rock by acid, these hollow spines are almost never found intact," explains Grant. "We think the spines may have served as 'snowshoes' to keep the brachiopods from sinking into the mud on the ocean floor."

Unlike clams or other shellfish that can move around and adjust to their environment, most brachiopods stayed put all their lives, either anchored in the mud or attached to rock or some other shells. Today, spiny brachiopods have died out, and brachiopods in general are no longer as varied as they once were: only about 300 species are in existence, compared to the almost 1,500 different ones known to have lived during the Permian. Why? Does their lack of mobility have anything to do with this?

"That's one thing we're trying to find out," says Grant, "particularly since the history of brachiopods down through the ages involves deep questions of ecology, evolution, and geology. The brachiopod record documents important phases in the history of life on earth, and we want to know that record as thoroughly as possible."

Movable Seas & Oceans

VIII. Digs

Reconstructing the spirit (and social organization, economy, history, and health) of societies past. The messages lie in the medium whether it is sand, tundra, or compacted soil covering an old graveyard. The bones, the pots, the very walls of houses and palaces are alive with information.

Excavating Hell

Sistan Desert, Afghanistan

Southwestern Afghanistan's Sistan Desert, a 5,000-square-mile area the Afghans know as *Dasht-i Jehanum,* "Desert of Hell," can legitimately lay claim to being one of the most unattractive, inhospitable, and odious places on earth. Vast desert lands form only one aspect of its forbidding desolation. Winter temperatures fall as low as 0 degrees F., and in summer rise as high as 130 degrees F.

From late May until late September the *bad-i-sad-o-bist ruz,* "wind of 120 days," holds the region in a paralyzing grasp. Gales howl day and night, raising the powdery soil in a perpetual dust storm. Spells of 80-mile wind velocity are not uncommon, and a high of 120 miles per hour has been recorded by at least one traveler. Then, as winds abate, clouds of stinging flies, yellow jackets, sand fleas, and mosquitoes arise. Poisonous snakes and spiders abound. Foxes and jackals scour the landscape. Outbreaks of malaria, cholera, and even bubonic plague take their dismal toll.

A 15th-century skull, discovered in a cemetery laid bare by continuous winds, lies on the compacted silt of Afghanistan's Sistan Desert. At right, a dust storm partially blurs ruins at Sar-O-Tar.

Sistan ruins, above and right, have
only partially withstood the elements,
and the next 50 years may prove even
more destructive. The structures
above stand in the inner city of
Sar-O-Tar, while the ruin to the right
was once an outlying manor house.

The Magnificent Foragers

A Baluch workman, right, excavates amidst the Sistan's harsh conditions.

Yet Smithsonian archeologist William B. Trousdale, who regularly explores this stretch of sand and solitude, has a deeper, happier view of the place. "Once this region welcomed man," he will tell you. For, from at least the sixth century B.C. until the 15th century A.D., hundreds of thousands of people lived here, supported by such agricultural plenty that geographers called the Sistan the "granary of the east." Today, vast sand dunes have buried most of the remains of this civilization, covering manor houses, villages, dozens of palaces, temples, forts, vast walled compounds, and at least three cities a square mile in extent. The most spectacular ruins belong to the Islamic period, particularly to the Ghaznivid and Timurid empires, which ruled much of the Iranian world during the 11th and 15th centuries.

"Because of the untouched isolation of its ruins," Dr. Trousdale claims, "it is the best place in the world, and perhaps the only place, to study unaltered 15th-century sacred and secular Islamic architecture. But these structures are rapidly succumbing to the elements. Within 50 years, less than half of what is left today will stand."

Trousdale's excavations have revealed the key to Sistan's past prosperity—a sophisticated system of dams and canals that controlled and distributed the waters of Afghanistan's Helmand River. At Sar-O-Tar, the Sistan's great urban complex, water to supply the city and its neighboring farms traveled 50 miles through a huge, high-banked canal. Numerous smaller canals, some of them longer than the great trunk itself, watered the surrounding region.

"Our archeological team mapped Sar-O-Tar's canals, discovered how the ancient gravity-feed irrigation system had worked, and identified crops that the canal water had nurtured," Trousdale explained recently. "We found that the city was planned with walled suburbs, 80-foot-high inner walls, moats, a circular administrative area, and an ornate citadel palace of soaring arches, grand courtyards, and pools. Sistan's complex system of water distribution dates back to the third or second millennium B.C.," he went on. "Neither the opening nor the closing dates for this unknown civilization have been determined yet, but we know that in addition to being superb engineers, these people manufactured elegant stone weapons and fine polychrome pottery."

The Smithsonian's expedition camp at Sar-O-Tar, above, is set up outside the ancient city walls. Archeologist William Trousdale, below, examines an inscribed 15th-century tomb tile in a mausoleum in the Sar-O-Tar area.

The first known period of sanding and abandonment took place in Sistan before the first century B.C. From the first century B.C. to the third century A.D. another occupation occurred, documented through coins and storage jars stamped with the insignia of the third-century Crown Prince Shapur and other rulers. Then, until the ninth century A.D., the Sistan appears to have been deserted again.

The Sistan's last period of occupation began in the ninth century A.D. Archeological discoveries by Trousdale prove that Genghis Khan's hordes sacked Sar-O-Tar in the 13th century A.D., ending two centuries of prosperity. Trousdale found evidence of this decline, as well as of a revival that came a century later under the rule of Tamerlane.

Clusters of magnificent 15th-century Islamic manor houses, some of more than 60 rooms, were built in the century of peace and creativity following Tamerlane's conquest. The ruins of these estates are scattered now throughout the desert along with ornate mausoleums containing tombs decorated with painted and inscribed tiles testifying to the high status of residents of the manor houses.

Another political decline followed a century later. The canal system broke down and sand blew out of ancient lake beds, smothering the region. These sand dunes, incidentally, may well be the fastest moving in the world. Fifty to 100 feet high, ripped by strong winds, they sometimes travel a foot a day. Today, only a nomadic tribe, the Baluch, and a few Afghan soldiers stationed at small military posts remain in the Sistan.

Trousdale's studies of the region suggest, however, that the dunes are not a permanent plague. "The periodic sanding of the Sistan may be a cyclical phenomenon," he guesses. "Twice after periods of desolation and emptiness lasting from 600 to 1,000 or more years, the region cleared itself of sand and was reinhabited by man. It is not beyond the realm of possibility that some day Sistan may flourish again and that the study we have made of the area's ancient and contemporary hydrology and agriculture may somehow aid in this rehabilitation."

The Magnificent Foragers

Layer Upon Layer

Koliktalik, Labrador

Labrador lies at the southeastern edge of the Arctic world, and, for thousands of years, maritime Eskimo and northern Indian hunting parties roamed its coastline and interior. When winter ice locked in the shore they would move to rocky islands off the coast where they could easily walk or sled out to the edge of the landfast ice to harpoon the whales, seals, walrus, bears, and birds that congregated near the open water.

For Smithsonian archeologist William W. Fitzhugh, these and other seasonal camps form part of a long-term investigation of the culture and history of the various Indian and Eskimo groups that have inhabited the Labrador coast since 5000 B.C.

One of Dr. Fitzhugh's many finds was the remains of an ancient sod house occupied about 400 A.D., possibly an early "duplex" inhabited by two Dorset Eskimo families on the island of Koliktalik, 30 miles east of the mainland town of Nain. Excavations revealed that the Eskimos had constructed the house by digging a rectangular cavity in the ground. Over this they erected a framework of spruce poles, which they banked with soil at a lower level and covered with caribou skins the rest of the way up. A smoke hole in the sloped roof provided ventilation. The doorway, a mere crawl passage, represented a traditional heat-saving design for Arctic winter living. Inside the doorway, probably once covered with skins, sat platforms to the right and left on which the families worked and slept. Two firepits lay along a central corridor, probably kept lighted to warm the house. A slightly raised kitchen area was bordered at the back by several large rock slabs. Here the archeologists found animal bones and grease, a small oval soapstone lamp, and fragments of large rectangular soapstone cooking pots.

Every winter the Eskimo group reoccupied the house after a summer spent hunting and fishing on the mainland. They "cleaned" the floor by laying down a fresh carpet of mossy sod. This practice afforded an unusual archeological slice of life.

Archeological crew lands on Nanuktak, or White Bear Island. Waterborne surveys are a necessity for research on the Labrador coast.

Smithsonian research vessel Tunuyak in Ramah Bay, site of ancient stone quarries used by Labrador Eskimos and Indians for the past 7,000 years. Abandoned only in the 20th century, its now-silent quarries are littered with broken tools. Below, crew members collect water for the base camp.

"We've found 15 sod layers, perhaps indicating that the house was occupied for 15 winters," Fitzhugh explained not long ago. "What we recover from between each layer encapsulates the history of one winter in the life of two Eskimo families. This gives a precise time frame in which to view the data we recover.

"And because we know how long the house was occupied, we've been able to get an idea of their total food intake over a 15-year period. We examined a garbage heap where they had discarded the bones of the arctic hare, fox, birds, seal, and other sea mammals they ate. We also took a look at their broken and used-up tools: harpoon points, stone knives, and scrapers. From these we learned something of what and how much they manufactured, how often they resharpened these tools, and how long the tools lasted."

Since many of his finds consist of stone tools, Fitzhugh is documenting the geological origin of the stone used by the Eskimos and Indians. Most of it came from several hundred miles away, probably through trade. One of his long-term goals is to locate native quarries along the coast, which will make it possible to trace the movement of the raw stone, some of which was traded as far south as Florida.

On the island of Nukasusutok, Fitzhugh found the remains of one of the earliest Eskimo houses ever discovered in Labrador, dating back about 3,000 years. Nearby was an Indian site 5,000 years old.

Excavations at Koliktalik, top, revealed evidence of a Dorset Eskimo winter house. Lichen-encrusted cobblestone figure adorns hilltop on Natsatuk Island near Koliktalik. Probably of Eskimo origin, the figure's age remains unknown.

"Fossil sea mammal blubber in the soil indicates that these Maritime Archaic Indians hunted seals, walrus, and maybe even whales under arctic conditions long before the Eskimos appeared on the scene," asserts Fitzhugh. "We know from other related sites in Newfoundland that they were using the 'Eskimo' toggling harpoon. It appears, in fact, that, contrary to earlier beliefs, the toggling harpoon may have been invented by Indians and passed on to the Eskimos."

When ice conditions permitted, Fitzhugh's party explored along the northern coast of Labrador. Sheer cliffs and mountains rise dramatically directly from the sea to elevations as high as 5,000 feet. Numerous fiords penetrate the Torngat mountains for distances as great as 30 miles inland. Here the archeologists saw caribou, seals, whales, ducks, and geese. Even the magnificent northern white bear, called "Nanook" by the Eskimos, appeared on a number of occasions. The names themselves of the fiord systems—Napartok, Saglek, Nachvak, Komaktorvik, and Kangalaksiorvik—go back to times long before the white man came to Labrador, when large groups of Eskimos hunted, fished, and lived along the coast.

Prehistoric and historic Labrador Eskimo hunting camps, large sod houses, and winter villages were discovered nearly every time Fitzhugh's crew members stepped ashore. Unlike in other areas of Labrador, some of the Dorset sod houses tend to resemble Eskimo dwellings. Fitzhugh believes that later Thule Eskimo immigrants to Labrador may well have adopted Dorset house types, which then became the most distinguishing feature of their culture in the eastern Arctic.

"We discovered many of these sites none too soon," says Fitzhugh. "It was sad to discover that maritime archeological sites in northern Labrador are being badly damaged by sea erosion. The land seems to be subsiding in this northern region. Many of the earliest Indian and Eskimo sites are already gone, and serious damage is occurring to traces of the Dorset, Thule, and historic Eskimo cultures."

Frankincense and Myrrh

Tel Jemmeh, Israel

No one lives at this desert outpost of the western Negev today, but it served as an economically and strategically important crossroads in the time of Solomon and Sheba, King Esarhaddon of Assyria, and Alexander the Great. Vital spice routes crossed here in the second and first centuries B.C. and connected the five ancient kingdoms of southern Arabia with the Mediterranean. While tracing those ancient camel-caravan highways, Smithsonian archeologist Gus Van Beek became intrigued by the successive tides of history that have swept over Tel Jemmeh, a high mound commanding an outlook on the ancient border between Old Palestine and Egypt.

A stroke of fortune—much like the blessing of oil in modern times—had given Arabian rulers a monopoly grip on the production and distribution of frankincense and myrrh. Only in the climate and soil of southern Arabia and across the Gulf of Aden on the African mainland, in what is now Somalia, could the scrubby trees that exude these gum resins be grown.

Dried and processed into powdery cakes, the two substances were in enormous demand throughout the Greek and Roman empires as well as in Mesopotamia, India, and Africa. Myrrh was blended into perfumes, cosmetics, and medicines, while frankincense was burned in temples and on funeral pyres as an offering to the gods. (It also served to disguise the odor of burning bodies.) The domes-

Workers excavate a large mud-brick silo on Tel Jemmeh, a hill overlooking the ancient border between Old Palestine and Egypt. The three pottery samples, opposite, were discovered inside the silo and date from the third century B.C.

The Magnificent Foragers

tication of the camel, a major transportation breakthrough, made
it possible for the Arabians to transport the two spices overland.
Spice-laden caravans could travel the 1,500 miles from southern
Arabia to Gaza on the Mediterranean in 65 days, making the last
of their overnight stops at Tel Jemmeh.

Dr. Van Beek concluded that Jemmeh appeared the most
promising site to dig for evidence of trade goods transported during
Biblical times. He was right. Initial exploration yielded pottery
sherds and other cultural remains from as deep as 45 feet below the
hilltop. The finds indicated that the hill was successively occupied
for 1,600 years. This habitation began about 1800 B.C. in the late
Bronze Age and lasted until 200 B.C. An earlier, limited settlement
had occurred in about 3000 B.C.

Tel Jemmeh had already given up tantalizing fragments of history. The famous Egyptologist Sir Flinders Petrie had dug there in
1927, uncovering ruins that included a group of 10 large mud-brick
silos with high conical domes. Constructed in the fourth to second
centuries B.C., when Palestine was part of the Egyptian Empire, the
silos provide evidence of a conservation project conducted by the
local people against the uncertainty of the Negev's periodic droughts.
Archeological science was relatively primitive in the 1920s, however, and Petrie's hasty investigation fell far short of solving the puzzles of the silos and other mysteries of Tel Jemmeh's past.

Selecting the highest point of the Tel to excavate, Van Beek
and his helpers hit an eleventh mud brick silo immediately below the
surface. It had collapsed, but walls still stood 7 feet high. Given an
opportunity to study one of Tel Jemmeh's silos in a modern, scientific light, they began carefully to sift through the layers of debris,
finding the ruins of a warehouse attached to the silo. Grain pots had
been stored there, and Van Beek believes the warehouse and the silo
fell simultaneously—perhaps during an earthquake.

Southern Arabic inscriptions have thus far been discovered on
the pottery, including the monogram of a caravan merchant named
Abum, whose inscriptions had previously been found at Marib and
Ma'in in north Yemen, and at Al-Ula in northern Saudi Arabia, all
towns on the ancient incense route. This tends to support Van

A palace bowl, top left, dating from the seventh century B.C., was excavated in one of six large rooms in an Assyrian building, right. Ron Gardiner, below, one of the supervisors in the Smithsonian Institution's Excavations in Israel, a volunteer program, excavates a deposit of eighth-century B.C. artifacts on Tel Jemmeh.

Beek's theory that the primary caravan route for the distribution of frankincense and myrrh reached Tel Jemmeh en route to Gaza.

Currently reconstructing a political, economic, and cultural history of the site across thousands of years, Van Beek uses the ancient fortification systems of successive occupations as an archeological index: 13th-century (Canaanite), 10th- and 8th-century (Philistine), and 7th-century (Assyrian). A ceramic kiln he has found which dates to the 12th century B.C. may shed new light on industry in the ancient town as well as add to our knowledge of early technology.

A great cobblestone courtyard with a plastered pool, probably remnants of a palace compound of the Canaanite occupation, is also under study. Most important of all, however, is a large six-room Assyrian building of the 7th century B.C. uncovered by Van Beek. Extensive portions of barrel-vaulted ceilings in mud brick are preserved in five of the rooms. It is the earliest example of this kind of architecture ever found in Israel, and predates Roman times. Moreover, it may be the earliest example of keystone-shaped voussoirs in vaults or arches in the history of world architecture. Iron spearpoints as well as the remains of fine imported pottery of the sort used in royal Assyrian households were discovered in the building's basement.

Ancient Assyrian annals of King Esarhaddon mention his conquest in 679 B.C. of a town known as Arsa on the Brook of Egypt, the traditional border between Palestine and Egypt. Van Beek believes that this was Tel Jemmeh and that the Assyrians rebuilt the town as a provincial center and military outpost, run by a military governor who probably lived in the building now being excavated.

"I like to think of the building as it must have looked when the Assyrians were here, with the governor relaxing in a chair on the roof at the end of day—perhaps with a glass of wine—and looking out over the fields toward the Mediterranean in the distance," Van Beek muses. "He was more than 500 miles from home and probably a little homesick. But as he watched one of the beautiful red sunsets over the distant sea, he must have thought that this lonely outpost wasn't such a bad place in which to serve after all."

The Magnificent Foragers

Jordanian Tombs

Bab edh-Dhra, Jordan

On Jordan's barren and eroded southeastern plain near the Dead Sea stretches a vast cemetery containing thousands of skeletons of people who lived in the Early Bronze Age, four to five thousand years ago.

In 1977, Smithsonian anthropologist Donald J. Ortner was invited by a team of scientists under the aegis of the American Schools of Oriental Research in Cambridge, Massachusetts, to coordinate excavations of the cemetery. It is located just to the south of the ruin of a large fortified town which the scientific team has under study and which is believed by some scholars to be associated with the ancient Biblical city of Sodom. Today the site of the ruin and the cemetery is called Bab edh-Dhra.

The earliest burials in this cemetery predate the town, going back to a time (3200-3000 B.C.) when the people in the area were apparently nomadic tribesmen, using the site primarily as a ceremonial center. Periodically the tribesmen brought the bones of their dead back to Bab edh-Dhra and placed them in unusual shaft tombs they had prepared earlier.

Most of the 4-foot-in-diameter shafts were dug to a depth of about 8 feet. Near the bottom of each shaft the tomb makers used stone tools to carve domed chambers into the soft lime clay, or marl. Up to five chambers were cut per shaft, each one about 6½ feet in diameter and 3 feet high at the center.

Crusader Castle in Kerak, Jordan, destroyed by Saladin in the 12th century A.D., served as temporary headquarters and base of operations for the team excavating an ancient cemetery at Bab edh-Dhra.

Digs

The tribesmen typically arranged the skulls of their dead with obvious care, setting them on a reed mat to the left of the tomb chamber. The long bones were placed in a pile in the center, and pottery vessels around the chamber's periphery, though also often concentrated to the right. In addition to pottery vessels, tomb offerings included stone mace heads, wooden shafts and bowls, and unfired clay female figurines. Once the skeletons had been placed in a chamber, it was sealed off by a wall of stone. After all the chambers were occupied, the shaft would be filled in, either intentionally or by neglect and natural erosion.

The existence of these shaft tombs had been known for several years, and many had been previously excavated by other scholars. Unfortunately, the skeletons that had been found tended to be in poor condition. The major purpose of Dr. Ortner's excavation was to recover a group of skeletons for subsequent research at the Smithsonian so he could attempt to clarify the relationship of the Early Bronze Age people at Bab edh-Dhra to other Near Eastern populations.

Ortner's team located the shafts by removing surface soil and gravel a few centimeters at a time, forming a trench 2 meters wide and 5 to 10 meters long. The outline of the shaft then became apparent because of slight differences in color and texture of the shaft fill. At this stage, skilled Jordanian technicians took over, carefully removing the fill down to the bottom of the shaft. Then the stones blocking the entryway to the burial chambers were removed so that Ortner could

A researcher, below left, peers into a tomb containing untouched pottery and bones at the bottom of an 8-foot shaft. Physical anthropologist Donald Ortner, below right, greets Sami Rabidi, a representative of the Jordanian Department of Antiquities who discovered the charnel house in which he stands. These mud-brick houses were used for surface, rather than underground, burial.

The Magnificent Foragers

The documentary photograph of shaft tomb findings, top, shows a central bone pile, a collection of skulls in the bottom left corner, and both broken and intact pottery. The jagged, time-worn hole in the rear of the tomb leads to a neighboring chamber. The pots shown in lower photo formerly lay on the tomb's floor; as silt oozed into the tomb over thousands of years, they gradually floated to the ceiling.

look in. Kneeling and shining a flashlight ahead of him into the blackness of the tombs, Ortner said he felt as though he had "rolled back 5,000 years of time."

More than half the tombs he opened were in an almost perfect state of preservation. Because of the adherent quality of the marl, which has the consistency of damp chalk, these chambers were structurally fairly strong. In the remainder, however, surface erosion during the rainy seasons had washed into the chambers and gradually filled them.

When the chambers were opened, Ortner took an initial visual survey, though often some cleaning had to be done so that the skeletons and tomb gifts would be completely exposed. This was necessary for documentary photography and the careful architectural drawings needed for subsequent research. Since subtle details in placement of skeletons or pottery may be significant, Ortner kept a record of the precise location of the tomb contents.

"Although most of the chambers contained disarticulated bones," he says, "one chamber was unique. We found skeletons of individuals in it who had been placed in the tomb at the time of death. There was a complete skeleton of an adult man about 40 years of age lying on his right side in a slightly flexed position. His hands had been placed together. Apparently the body was clothed at the time of burial, since we found the badly decayed remains of cloth associated with it. We also found the complete, disarticulated skeletons of two children and the skull of an infant in the tomb. Apparently, the two children had been placed in the chamber sometime before the adult, since their bones had already been scattered when the adult was interred."

The tombs and burial practices give evidence of considerable pains taken on behalf of the deceased, suggesting the importance of the dead in the culture of these people. According to Ortner, the act of cutting the shafts and tombs must have taken a great amount of work. It took his crews, with technician and three to five workmen, from four to six days to reach the chamber. The pottery, which was much finer in quality than the everyday variety used by the people, leads to the same conclusion. In some of the pots, moreover, the excavators found grape seeds and a residue that chemical analysis indicates may be dried milk. This suggests that food offerings may also have been placed in the tombs.

"The excavations at Bab edh-Dhra illustrate the changing cultural patterns of the area," says Ortner. "Toward the end of the shaft tomb period a more sedentary, town-oriented society replaced the nomadic-pastoral culture, at least for a time. In this new urban phase, customs changed: Bab edh-Dhra's people stopped burying their dead in underground tombs and shifted instead to surface burials in mud-brick charnel houses. We excavated a number of these. I hope to learn from the studies of the skeletal material we brought back to the Smithsonian whether the change from a nomadic life to a sedentary one was the result of a single culture changing through time, or the result of the influence of a more urban people who conquered the indigenous nomadic tribesmen of Bab edh-Dhra. Hopefully, the skeletons will tell us."

Relative Longevity

Maryland; Ecuador

Why do some people live longer than others? Evidence suggests a combination of factors, ranging from a nourishing diet and a healthful climate to a favorable genetic structure. Whatever the causes, it is known that longevity and other vital statistics vary considerably from one population to another. The Smithsonian's Douglas H. Ubelaker investigates these variances among prehistoric people in both North and South America.

Late Woodland Indians who lived throughout the mid-Atlantic region buried their dead in ossuaries, large communal graves. When excavated properly, their contents make it possible to draw population profiles of these prehistoric Indians. One such ossuary was turned up in 1971 by a Maryland man who was digging a post hole on a farm in the southern part of the state: his auger struck bone. Learning of it, Dr. Ubelaker soon arrived on the scene and began to excavate. He uncovered an ossuary that contained the remains of at least 188 Indian skeletons of various ages, buried about 1500 A.D. After estimating the age at death for each individual, he determined that the population had a high infant mortality rate (30 percent died before the age of 5), a life expectancy at birth of about 23 years, and an average adult age at death of about 36 years.

According to Ubelaker, many of the young adults in this society may have been killed by warfare or succumbed as a result of occasional drastic food shortages related to climate. He points out, however, that evidence has been put forward that life expectancy at birth in Europe during the 16th century may also have been as low as 20 to 25 years. If this is correct, the Indians in Maryland were no worse off than Europeans in terms of life expectancy.

A distinctly different picture emerged at a prehistoric cemetery on the southern coast of Ecuador, also excavated and analyzed by

Excavations of prehistoric cemeteries on the south coast of Ecuador, below, make it possible to draw population profiles of the area. The cemeteries date from around 1500 A.D. Amateur archeologist Earl Lubensky, below left, examines material from the ancient Ecuadorian cemetery he discovered, while Douglas Ubelaker, below right, studies extended skeletons at the same site.

The Magnificent Foragers

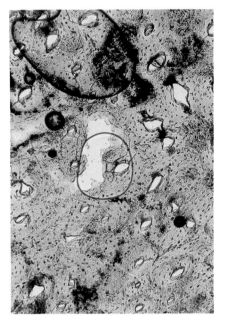

Above left, a microscopic cross section of a human femur. By using a formula based on the percentages of various structures in view, scientists can estimate age at death. Dr. Ubelaker compares slide to original bone, above right.

Ubelaker. There he found that the Indians who made it to adulthood lived significantly longer than the Maryland Indians. Also dating from about 1500 A.D., the Ecuadorian cemetery consists of 43 large, buried ceramic jars, each containing an average of nine individuals of all ages. Ubelaker theorizes that the urns represent family burials since all ages are present and since many of the skeletons within single urns share physical characteristics. Careful analysis of ages at death revealed that, like the southern Maryland group, this population had a high child mortality rate—38 percent of the population died before the age of 5. The life expectancy at birth was about 23 years, but the average adult age at death was 41 years, five years greater than that of the Maryland population.

Contemporary residents in this area, which is situated just south of the Equator and cooled by the Humboldt and Niño currents, suffer from few respiratory ailments. It can probably be assumed that their forebears enjoyed a similar freedom. And where the Indians of Maryland appear to have been affected by periodic food shortages, the 16th-century Ecuadorian group most likely could depend on a steady supply of maize, fresh- and salt-water fish, and other aquatic foods. Interestingly, the Ecuadorian cemetery is only about 50 miles northwest of the village of Vilcabamba, where in 1971 scientists suggested that there might be an unusually high percentage of very old persons.

Several factors probably gave the 16th-century Ecuadorians a better chance to live longer. "For one," notes Ubelaker, "there is no evidence of warfare in this society. No weapons were found in the excavations, nor any sign of bone damage from projectile points or blades. There also seemed to be a lack of disease. The skeletons were surprisingly free of pathology. And many of the diseases that historically lower life expectancy—syphilis, malaria, measles, mumps, and smallpox—were either nonexistent or were not severe problems until after the Spanish arrived. Because of the remote area in which they lived, this group of Indians apparently remained isolated from the Spaniards until long after the colonization of Ecuador."

Problems of the Ancient Greeks

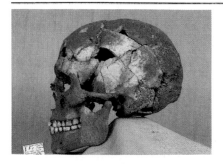

The skull of a Mesolithic male dating from before 7500 B.C., above, was excavated at Franchthi Cave, 15 miles from Lerna. The skulls below illustrate the sequence of age changes in Middle Bronze Age skulls. Bottom row: the skull of an infant, far left, and those of two progressively older children. Top row: the skull of a young adult, left; a middle-aged adult, center, and an old adult, right.

Lerna

Prehistoric man had ecological problems, too.

According to Smithsonian physical anthropologist J. Lawrence Angel, some of man's earliest efforts to manage the environment for his own economic well-being produced side effects deleterious to his health. Dr. Angel recently studied a prehistoric Bronze Age archeological site at Lerna, now the modern Greek village of Myloi on the Plain of Argos. Excavations there by the American School of Classical Studies at Athens have shown that it was continuously occupied as far back as 6000 B.C., when its favorable setting on the western shore of the Bay of Argos attracted Neolithic fishermen and farmers. The most significant find dated from the Middle Bronze Period (2000-1600 B.C.): archeological teams found 235 graves under floors and courtyards, covering a span of 25 generations and marking the first unearthing of such a large prehistoric family burial ground.

Because of his interest in social biological changes in the eastern Mediterranean from 10,000 B.C. to the present, Angel during the past two decades has examined almost every human skeleton unearthed by archeologists throughout Greece and the Aegean. He was therefore intent on looking at the Lerna population.

By measuring skeletal remains—modern or ancient—Angel can discern a person's health, body build, growth, and in many cases his occupation, all of which are determined by diet, climate, living habits,

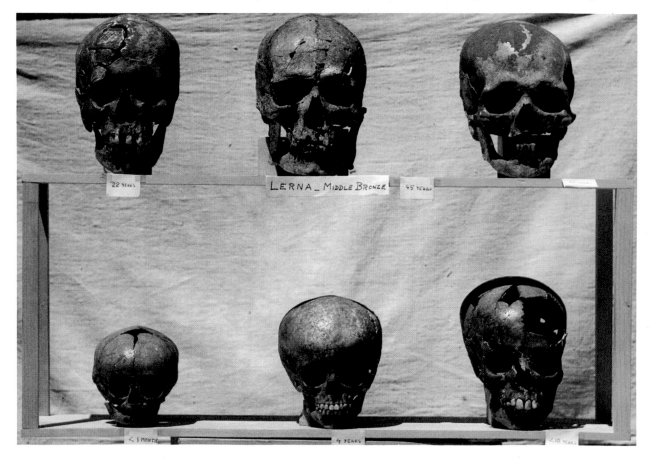

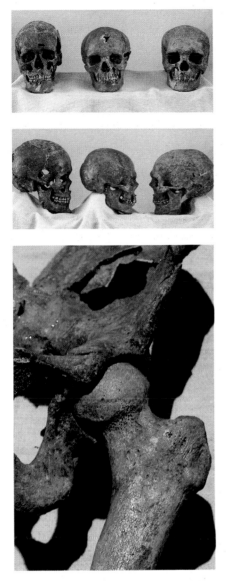

and heredity. Skulls and bones also give clues to genetic relationships, evolutionary changes, and migrations.

The demographic profile Angel constructed shows that adults in Lerna (which he estimates had about 800 persons living in it during the Middle Bronze Age) had an average life expectancy of 34 years—37 for men, 31 for women. The average woman bore about 5 children, 2.2 of whom grew to adulthood (15 years of age). Because of that birthrate the population was increasing, doubling every seven to ten generations, a surprising figure considering the handicapping diseases that Angel found also afflicted the community.

Malaria, in particular, brought about a crippling impact. The changeover from a hunting to a farming culture, which began in the Mediterranean between 9000 and 6500 B.C. with the disappearance of the big game herds, drew early farming populations to sites like Lerna, where the soil was soft and the forests not established. But marshy areas favorable to farming also favored the Anopheles mosquito. Malaria, especially the type known as *falciparum* malaria, ran rampant, and its effects are plainly visible in the skulls Angel examined.

Angel believes that physical debilitation caused by the malaria and other endemic diseases plagued and weakened the population of Lerna for most of its prehistoric period. The average stature of the Lerna men was 5 feet, 5½ inches, and the women only 5 feet, ¼ inch. Malaria may also have made the people of Lerna vulnerable to invading bands from the north and east that swept across Greece between 3000 and 2000 B.C. These conquerors remained in Lerna and intermarried with the local people, creating somewhat of a "melting pot" as evidenced in the heterogeneous skulls that show a blending with immigrant stock.

Malaria was not the only health problem ravaging the residents of Lerna. Angel found a heavy incidence of arthritis and dental disease (25 percent of the teeth of the Lerna population were diseased or missing). He is convinced that, as with the malaria, the dental trouble resulted largely from the changeover to farming. A diet of proportionately less meat and of cereals containing fewer trace minerals apparently proved adequate for energy but less so for growth.

Sleeping in unheated houses probably caused the high incidence of arthritis. Fully 75 percent of Lerna's adult men suffered in some degree from arthritic backs and necks, a dismal record surpassed in the annals of pathology only by modern Eskimos.

Two separate shots of the same three men—frontals above, profiles below—show racial mixture in Lerna around 2000 B.C. The right-hand skull is representative of pre-Greek natives, while the other two belong to Greek-speaking intruders to the area. The front view of a hip bone, below, displays ligament stress, probably caused by climbing mountains.

*   *   *

Another Smithsonian physical anthropologist, Lucile St. Hoyme, uses the Institution's collection of 20,000 skulls to study the causes of tooth decay and dental diseases. While examining the lower jaw of an Indian woman who had lived between 900 and 1200 A.D. near the present St. Louis, Missouri, area, Dr. St. Hoyme and a colleague discovered evidence of the first known dental fillings. A prehistoric Indian "dentist" had drilled into cavities in the woman's right canine and first premolar, then neatly filled them with a cement-like substance.

"We know that Mexicans and others of the period had fancy jade inlays for cosmetic reasons," says St. Hoyme, "but this is the first evidence of a filling for therapeutic reasons."

IX. The Ice Ages & Before

Paleobiologists try to interpret and understand fossil organisms and their relationship with the environment. They do so by reading the sedimentary rock record as a history book, each layer being a page, and the bottom layers being the earliest chapters. Fossils are like words on those pages, explaining the biological events of each time. And sometimes they even attract archeologists.

Paleo-Indian Hunters in America

Northeast Colorado

The questions of when prehistoric man first arrived in the Americas, by which route, and whether he came at one time or in successive waves, continue to stir scholarly debate among archeologists. Who were these first Americans, and what was the chronology of their movement southward?

Datable artifacts and bones at more than a half dozen sites in the North American heartland long supported the view that Paleo-Indian hunters of woolly mammoth and other big game first made their way into the New World from Asia between 13,000 and 11,500 years ago. But clues now are turning up that cloud the accuracy of this picture. The Smithsonian's Dennis Stanford, a specialist on Paleo-Indians, has examined a complex of sites in northeastern Colorado that he believes will alter our perception of man's earliest history in America.

"None of our finds would have been possible if Robert B. Jones had not set out in 1972 to dig an irrigation system in a field on his ranch near Wray, Colorado," Dr. Stanford recalls. "In the course of bulldozing, Jones chanced upon some bones and flint points. He promptly called an anthropologist, Jack Miller, to look over the site. When Miller saw that it contained bison bones and a large number of Paleo-Indian flint and bone tools, he notified the Smithsonian." Shortly after, Stanford took in a crew to excavate.

Some 41,000 bones from 300 Bison antiquus lie exposed at an archeological site in northeast Colorado. The bison were killed and butchered nearly 10,000 years ago.

The Ice Ages & Before

With the aid of poles cut with a stone adz, researchers turn over the half-butchered elephant to begin work on the other side.

An Archeological Butchering

The accidental death of a female elephant at a Boston zoo in January, 1978, provided Stanford an opportunity to conduct an unusual experiment in Pleistocene archeology. Viewing the elephant as a stand-in mammoth, he wanted to subject obsidian stone tool replicas—flaked and hafted onto wooden handles by a colleague, Dr. Robson Bonnichsen of the University of Maine—to real butchering, not only to compare the signs of wear produced on these new tools with those borne by actual Ice Age artifacts, but also to measure the amount of human exertion such butchering required.

Accordingly, after intense negotiations, the bulk of the carcass was transferred on a flat-bed truck to rural Virginia, where the Smithsonian's National Zoo operates a breeding farm. The weather stayed cold enough to keep the elephant frozen. On the appointed day, the experiments began.

First the researchers tested the difference in penetration between flint and bone spearpoints, hurling spears at the carcass with an atl-atl. The bone-tipped spears penetrated more deeply, presumably deep enough to bring down a living mammoth. Then they set to work in earnest on butchering: hacking, chopping, sawing. It quickly became apparent to a perspiring Stanford that the energy expended in butchering was "tremendous." But, contrary to expectation, the degree of wear produced on the tools proved negligible, casting doubt on conventional archeological explanations for the chipping and polishing often found on genuine prehistoric tools.

To test another hypothesis, the experimenters put aside their obsidian tools and, by smashing a leg bone with ordinary stones, produced bone fragments that resembled those found at ancient sites and flaked into sharp, effective knives. "It seems altogether possible," concluded Stanford, "that Pleistocene hunters could have killed and butchered a mammoth entirely without stone tools." The implication is that a lack of stone tools at a prehistoric site need not automatically mean a lack of human occupation.

Students butchering one of the elephant's feet with unmodified flakes of chert while a third records data on the work itself, including tool numbers and length of time in use.

A bone-flake knife cut the elephant meat as readily as a stone knife.

They unearthed the bone remains of nearly 300 big-horned bison (the extinct *Bison antiquus*) spread over an area 30 meters by 20 meters where they had been killed and butchered some 10,000 years ago by a band of as many as 50 Paleo-Indians. These ancient hunters had tossed the bones into piles, suggesting that the group had been organized into work parties by someone in charge. Each party apparently had specialized responsibilities for preparing meat cut from different sections of the dead animals. Also, hundreds of cutting and chopping tools, as well as stone projectile points, turned up at the site. The source of the stone has been traced to Plains areas in Wyoming, Nebraska, Kansas, Colorado, and the Texas panhandle. This raises the question of whether the stone arrived there through the union of several bands at the site, one band's roving travels, or regional trade.

In any event, Stanford now has evidence that one or more groups of Paleo-Indians wintered in northeastern Colorado, and on several occasions assembled to ambush bison at the Jones-Miller draw. "In the scenario I reconstruct," he adds, "the draw was filled with wind-blown snow. The hunters herded a group of bison cows, calves, and yearlings into the draw. As the trapped animals floundered around in the snow-drifts, hunters at the edge of the draw dispatched them with hand-launched spears. Work teams butchered the bison on the site. Snow chilled and kept the meat fresh until the job was done."

Many observers of the 19th-century Northern Plains Indians reported the use of a similar buffalo-hunting strategy and noted that hunts were highly ritualized affairs, with a distinct ceremonial character. Typically, a "medicine post" was set up at the center of the impoundment, and offerings for a successful kill placed around it. For several days preceding the hunt, the hunt chief held a religious ceremony outside the impoundment, burning incense on smudge fires.

"There was intriguing evidence that Paleo-Indians practiced these same rituals at the Jones-Miller site," says Stanford, who discovered a large post mold in the center of the draw, but so shallowly emplaced that it could have hardly served any purpose in the butchering operation. Near the mold, moreover, lay a flute-like drilled bone and an extremely tiny but complete projectile point. Both of these could have

The Ice Ages & Before

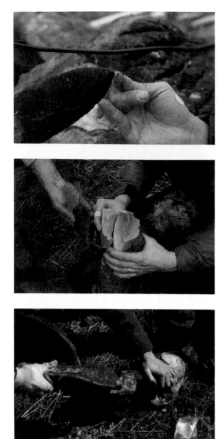

Tiny flakes chipped in the butchering process and visible at the end of the index finger are compared, top, to the edge from which they were broken. Small flakes of this nature have been found in excavated archeological sites. Middle: the manufacture of bone tools with the humerus serving as an anvil. Using a cobble, a striking platform was made on the spirally fractured humerus. From this prepared platform, bone flakes were struck and employed in butchering. One of the bone flakes is compared, bottom, to a nearly identical specimen found in the Canadian Arctic and possibly dating to more than 30,000 years ago.

fulfilled a ceremonial purpose. In addition, he located a hearth area west of the bone bed that contained red and yellow ochre, substances associated with ceremonial activities.

Then, as Stanford was packing up to return to the Smithsonian after his third summer of work at the Jones-Miller site, serendipity struck again. A construction worker who had been digging an irrigation pond at the Selby Ranch, 20 miles away, stopped by. His bulldozer blade had hit what looked to him like mammoth tusks, he explained. Stanford's crew drove to the ranch and confirmed the worker's guess: he had indeed found mammoth tusks and evidence of human occupation.

Stanford returned to Colorado to excavate, this time concentrating on the Selby site as well as on yet another one near the Dutton Ranch where the same bulldozer operator turned up more bones. At both sites Stanford uncovered successive layers of tools and bones, including the remains of mammoths, camels, giant bison, ground sloths, horses, lions, fox, bear, deer, and antelope. Geological analysis showed that between 10,000 and 20,000 years ago the sites had been the banks of small marshy lakes. The animals that grazed on the surrounding highlands had come for water and apparently been killed and butchered there by Paleo-Indian hunters.

The top level at the two sites contained thin, fluted stone projectile points, characteristic of the Clovis Paleo-Indian period between 11,200 and 11,500 years ago. Clovis tools have long been considered the earliest Paleo-Indian stone artifacts. Below these, however, and presumably older, lay a stratum with bone choppers and scrapers. Significantly, no stone projectile points or tools were found. A mammoth bone from this pre-Clovis level was carbon-dated as some 11,700 years old. Below the second level two more levels appeared that also contained bone tools and bones of butchered animals.

It is already clear that the two sites contain the most complete record yet discovered of Ice Age man in America, and that the two lowest levels indicate that man has been in the New World nearly 10,000 years earlier than the previously accepted date of 11,500 B.P. (Before Present).

The Magnificent Foragers

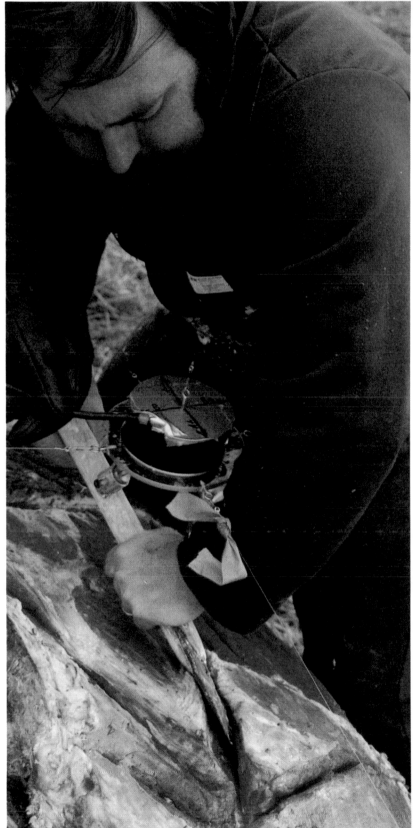

A femur was broken with a 21-pound boulder, producing spiral fractures and impact depressions identical to those found on mammoth bones from archeological sites. Right, Dennis Stanford cutting elephant meat with a hafted bifacially flaked knife. An apparatus for recording changes in the angle of the knife is suspended in front of Dr. Stanford and attached to the knife's handle.

Fossil Potpourri

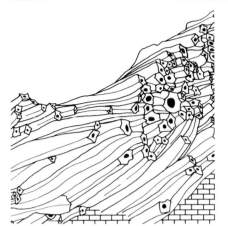

A schematic drawing of colonial rudists, an extinct reef-building mollusk studied by Museum paleontologist Erle Kauffman. At one time, 35 to 40 million years ago, rudists out-competed corals and became the world's primary reef builders. The illustration shows a cluster of Biradiolites jamaicensis, whose elongate shells leaned over and grew sideways, a common rudist strategy for generating the reef matrix. Lower drawing depicts a typical foraminifera belonging to the genus Elphidium, magnified 120 times.

The Briny Deep

Whether microscopic or gigantic, fossil organisms give us a glimpse of the earliest and most primitive life forms. "Most animals and plants that have lived on earth are extinct, and their study is the business of paleobiology," explains Erle G. Kauffman, an expert on fossil mollusks. "Our collections tell us much about the history of the earth over the last three billion years—the age of the first fossil. Over this span there have been many millions of time periods in which the earth was covered with plants and animals very different from those of today," Dr. Kauffman continues. "No two of those time periods were alike, as the evolution of life has been progressive, not repetitive. Documenting the various forms of life that flourished during those periods is a major part of our work. It is a task far from finished. It is still not unusual to find a completely new flora and fauna that lived in the remote past."

*　　　*　　　*

A handful of mud off the sea bottom usually contains thousands of living shelled protozoans called foraminifera, each the size of a sand grain. They abound in marine waters everywhere, and most sedimentary rocks contain great numbers of fossil foraminifera. In this century these fossils have been collected and studied intensively by geologists who use them as guides to oil and gas.

Joseph Cushman, a Harvard geologist, was one of the first to realize their economic importance. He demonstrated that fossil foraminifera showed change through time as a result of evolution, and that they could be used to date rocks. His collection forms the nucleus of the Smithsonian's foraminifera collection, which now contains over a half million identified specimens, making it the largest in the world.

Ocean currents transport living planktonic foraminifera over hundreds of square miles. The Smithsonian's Richard Cifelli is charting the distribution and abundance of these tiny creatures which do not swim and must go where the currents take them. Typically, he drags a fine-meshed plankton net in the North Atlantic, separating the "forams" from the other animals in the net with a special furnace technique he has developed. In the furnace the soft, organic-tissued plankton burn up, leaving the shelly remains of foraminifera. When these are identified and counted, he can plot the species' distribution and relate it to ocean currents. By doing so he can see that different water masses in the North Atlantic are characterized by different species associations. "When you look at the fossil record you can't see the water anymore; but you can see the species' association, so you can make estimates of what the water masses were like," says Dr. Cifelli.

Martin A. Buzas also works on living foraminifera, which he says represent the key to the past. "Except for parasites harmful to man, protozoans have not been studied much by biologists," he points out. "Consequently, many geologists like myself have shifted to the study of the living animals because we need information about their ecology if we're going to interpret the fossil record we find in the rocks.

"I've been interested in ecological factors that bear on foraminif-

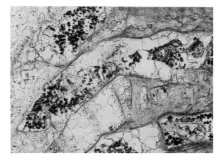

Photomicrographs of bryozoans, above and below. Paleobiologist Richard Boardman has developed miscroscopy techniques that enable him to study the incredibly fragile internal structure of bryozoan fossils, such as that of the specimen, top, that lived approximately 550 million years ago. The internal structure is revealed by concentrations of iron-rich granules in fragments of fossilized membranes. The basic funnel shape of the tentacles and guts of modern bryozoans, as seen in the specimen in the lower photomicrograph, can then be directly compared with remnants of soft tissues in fossils. Bottom right, living bryozoans, relatives to species shown in slides.

era distribution and abundance," continues Dr. Buzas. "Why do different foraminifera species live in different sediments, and why do their numbers vary? Until recently we sought answers to these questions by looking only at physical conditions such as temperature and salinity of water. But I'm getting data that indicate competition between different species of animals is equally important. I'm trying to find out what effect predation has on foraminifera populations and how large a role the foraminifera play as food for clams, crabs, fish, worms, and other marine animals."

*　　　*　　　*

Tiny underwater animals, bryozoans comprise a class of aquatic, mostly marine, invertebrates that reproduce by budding and ordinarily form permanently attached branched or mossy colonies. Their study is the domain of Smithsonian paleobiologists Richard Boardman and Alan Cheetham.

"Like fossil foraminifera, fossil bryozoans don't provide the whole story," says Dr. Boardman, "so we collect and study living ones to learn more about fossil colonies in which only skeletons are available."

Bryozoan colonies either encrust a submerged surface or grow erect, forming fan- or bush-like shapes. The larger colonies contain several million microscopic members, specialized to perform either feeding or reproductive functions. Living bryozoans are collected from the sea bottom by diving or dredging.

Dr. Cheetham's studies bring statistical and computer techniques into play. Measurements and statistical summaries of individual anatomical bryozoan components are necessary to understand the animals' growth and evolution, and only the most advanced mathematical methods make such summaries possible.

The Ice Ages & Before

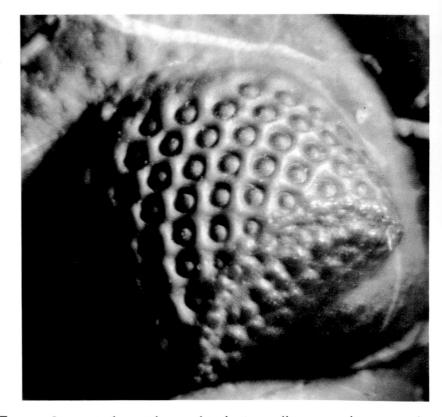

A photograph of a trilobite's calcified eye shows some of its 75 lenses, each of which measures about a half a millimeter in diameter. Bottom, two thin sections: the upper, a cross section; the lower, a close-up of one of the lenses. The small cross in its center is a result of the polarized light beamed through the section to study lens crystallography.

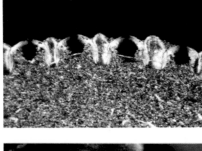

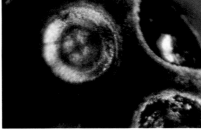

Lenses and special optical techniques allow researchers not only to look into the microscopic world, but to see into the past as well. Kenneth Towe investigates the shell structure of tiny foraminifera and other, larger marine creatures in the collections of fossils. Among these are trilobites, extinct sea-floor dwellers. Dating to the Cambrian period more than 500 million years ago, they were then one of the most abundant forms of life in the sea.

Not long ago, Dr. Towe, a geologist, decided to look more closely at the structure of fossil trilobite eyes. Surprisingly, he found that their eye lenses had not hardened as a result of post-mortem fossilization, but had been hard in life. Where most eyes—both ancient and modern—are soft tissue, and therefore perishable, trilobite eyes were formed of calcite, a limestone-like calcium carbonate. Remarkably, each lens of the multi-lensed eye turned out to be a single calcite crystal. Moreover, each lens is so precisely oriented crystallographically that it behaves optically as if it were made of glass. As a result, the trilobite could have seen clear images over a wide angle of vision.

"Calcite crystals produce a double image unless they are perfectly oriented to behave like glass," explains Towe. "These animals that lived on the bottom of the sea had, incredibly, evolved the precise correction to avoid a double image." One of the few living creatures today possessing such calcified eyes is the sow bug, though it has failed to solve the double-image problem. How the trilobite did remains an evolutionary mystery, concedes Towe, though he does offer a possible reason why calcified eyes did not become more widespread: "The difficulty with such a visual system," he says, "was that it couldn't adjust to distances. The depth of field was always locked in."

The Magnificent Foragers

Titanotheres in the Badlands

Wyoming

The Badlands, so aptly named by the Dakota Indians, were formed where clay, siltstone, and sandstone sedimentary formations eroded into pinnacled and ridged masses. Weathering out of the rock deposits in these strange-looking landscapes are the fossilized bones of titanotheres—large, heavy-bodied, rhinoceros-like horned and hoofed mammals that lived about 35 to 31 million years ago in the Oligocene Epoch. Living with the titanotheres in the warm forests that covered much of what is now the western United States were many other beasts, including collie-sized horses, 2-foot-high camels, and much smaller insectivores and rodents.

Today, after more than a century of scientific exploration in the badlands areas, scientists are still uncovering fresh data about these extinct animals. The Smithsonian's Robert Emry focuses on badlands outcrops of the 750- to 800-foot-thick White River Formation in central Wyoming. Oligocene fossil deposits in this area are interbedded with layers of volcanic ash, which can be dated by measuring the potassium and argon content. By studying the changes in structure of bone and teeth of various fossil animals from one level to the next, Dr. Emry can document the time it took for evolutionary change to occur in various animal lineages.

Searching for his study material, Emry systematically works his way back and forth along badlands slopes, his eyes glued to the ground. "In several summers' work I've brought back over 6,000 specimens," he states. "Often I'll first spot bone fragments that have eroded out and washed downhill, and then trace the fragments up the slope to find the rest of the bone in place in the outcrop. Sometimes it will be part of a titanothere or some other large animal, but much more often it will be bones of small rodents, insectivores, or other smaller animals. I usually find fragmentary parts of jaws with teeth, or even single teeth. Occasionally part of a skeleton turns up, but a complete skeleton is a rarity."

Emry is interested in the small fossils as well as the larger. Small animals are generally more closely attuned to particular habitats and have the advantage of telling us more about the local environmental conditions than the larger animals. Most of the older collections from the White River Formation badlands are biased toward the larger animals, however, since these are more easily found. Consequently, the composition of the area's animal population at different times in the Oligocene remains uncertain.

Occasionally fossil bones of small mammals, such as mouse-sized rodents and shrew-sized insectivores, are found concentrated in a small area, perhaps as many as 200 to 300 specimens in a few cubic feet of sediments. According to Emry, these probably represent nests or dens of small mammalian carnivores or owl roosts (fossil owls are found in the deposits). When this kind of concentration is found, Emry bags the sediment so that it can later be soaked in water and washed through screens. The residue, containing bones and teeth as well as the coarser sediments, is sorted in the laboratory by the use of heavy liquids. The fossil fragments, particularly teeth, are usually heavier than the rock particles. A chemical several times denser than water is

Collecting titanothere remains in White River Formation on Beaver Divide in central Wyoming, top. Close-up of lower jaws of titanothere, bottom, after being jacketed with plaster of Paris bandage on one side and turned over. The exposed side will also be jacketed before the specimen is moved to the Smithsonian.

The Ice Ages & Before

A Smithsonian technician in a badlands outcropping in central Wyoming's White River Formation makes a cast for the skeleton of a sabertoothed cat.

diluted just to the point that the rock fragments float on it while the fossil tooth and bone fragments sink.

"When a large fossil bone is found in the field," says Emry, "I dig a trench around it, exposing the specimen on a pedestal of rock. It is then encased in a cast made of burlap or gauze dipped in plaster of Paris—much like the casts doctors put on broken limbs—so that it can be transported back to the Museum without danger of being broken."

Emry regrets that good fossil material is increasingly difficult to find in certain areas, partly because of the greatly increased number of amateur and commercial collectors. "I make a distinction here between amateur and commercial collectors," he explains, "because they collect for different reasons. Most amateurs are quite professional. They collect because they are interested in the fossils themselves. Because of their interest they often learn as much as they can about the stratigraphy of the deposits and the kind of fossils to be found in them, and are able to identify their specimens. They usually know when they have found an unusual or scientifically important specimen and bring it to the attention of a professional vertebrate paleontologist. Many important specimens now in scientific institutional collections were found by amateurs who were interested not only in adding to their own collections, but in furthering the science of paleontology as well. Commercial collectors, on the other hand, are usually interested in fossils primarily as objects that can be sold. The fact that a fossil is a rare or even new species, and may be important scientifically, simply means that it can command a higher price.

"There's a ready market for fossils," he goes on. "People want them as curiosities, to put them on their mantelpieces or desks. Ironically, a scientifically important specimen may become a doorstop. I see fossils that I know came from where I collect in the Wyoming White River Formation listed in sales catalogs. Because of the commercial collectors, many landowners are now keenly aware that fossils have a commercial value. Some ranchers whom I've known for years still give me permission to look for fossils on their property, but I sometimes feel uneasy because whenever I find a specimen they want to know how much it is worth. I have to keep pointing out that scientists do not value specimens monetarily. To us, their value is in the scientific information they convey."

Therapsids

A drawing of therapsids, animals that lived 200 million years ago. Officially classified as reptiles, they possessed numerous mammalian skeletal traits, including, in many predacious species, dog-like canine and molar teeth. These animals are thought to represent intermediate stages in the evolution of mammals from reptiles.

South Africa

Nicholas Hotton III, curator of fossil amphibians and reptiles for the Smithsonian, was born in Upper Michigan, "in a part of the country having no lizards at all, very few snakes, and nothing in the geologic record from the Age of the Dinosaurs. So naturally," he adds wryly, "I became interested in the paleontology of reptiles."

His work since has led him into investigations of animals technically called therapsids, bizarre, mammal-like reptiles that lived 200 million years ago and form an intriguing intermediate link in the evolution of mammalian life from reptilian levels.

Therapsid remains are very scarce in North America. But sedimentary deposits containing them are known in India, China, Antarctica, and South Africa. For Dr. Hotton's purposes, South Africa has proved the most rewarding and convenient place to collect. The fossils are commonly found in river and deltaic sedimentary deposits, indicating that the animals lived primarily along the margins of rivers and streams. According to Hotton, dinosaurs eventually invaded these habitats and drove the therapsids into extinction. But before this happened, the therapsids gave rise to the first mammals.

"In general," says Hotton, "certain anatomical details can be seen in the therapsids I study that set them apart from earlier reptiles. The limb bones were longer, more slender, and pulled in under their body, instead of being spraddled out, turtle fashion. Trunks and tails became shortened, so that they no longer looked like lizards. This shortening reduced the tendency to move in an undulating, fish-like manner; limb motion had become more important than back motion. Changes in jaws and teeth enabled more efficient mastication so that digestion proceeded more rapidly. Mammal-like air passages evolved, and a few advanced forms may even have developed hair."

These changes probably enabled therapsids to cope with temperature fluctuations better than their contemporaries, Hotton believes. The ability to control temperatures gave the early mammals a greater tolerance over seasonal extremes than dinosaurs and may have been one important reason that they eventually replaced the dinosaurs as the dominant order on earth.

In particular, Hotton has been studying therapsid earbones. "Earbones are curious things to be useful in identifying an animal," he says. "But all earbones work the same way, transmitting vibrations to the inner ear. However, they appear to be structurally different in various species of therapsids. My work is all the more difficult because of the oddness of the animals," Hotton goes on. "You put therapsid bones together and you know how they must have fit, but the result just doesn't look plausible. The reason, I think, is that we have a gut-level familiarity with cats, dogs, horses, elephants, lizards, birds, and the like from having lived with them so long. But, obviously, we have no such backlog of experience with living therapsids because they have been extinct for some 200 million years. Even so, while they lived they could hear things—perhaps the footfalls of approaching predators—and, somehow or other, they made a living. It intrigues me to be able to put their bones together and try to interpret just how they managed to do so."

The Ice Ages & Before

Root of the Devonian

Francis Hueber found this fossil trunk fragment during his second trip to Australia. The striping and ridges on the trunk represent the remains of the wood structure.

Queensland, Australia

In late January 1975 in the cattle ranching country of northern Queensland, Australia, paleobotanist Francis Hueber pried out of a low sandstone ledge the fossilized remains of a 360-million-year-old Devonian plant. It was the best preserved specimen yet found of this Devonian genus, and for Dr. Hueber it represented a major breakthrough on a project begun 17 years earlier.

Back in 1958, he had collected four or five isolated fragments of that genus in New York's Catskill Mountains. Unfortunately, the fragments—oddly star-shaped in cross section—were not large or complete enough to indicate much about the plant. Still, Hueber was fascinated and wanted to find more. Because it was unlikely that additional specimens would soon turn up in New York, Australia seemed to him to be the best place to look. He knew a piece of the Devonian genus had been found in the 19th century at a site on the Fanning River in northern Queensland, though a description of the genus, classifying it as a herb, was not published until 1929.

More than nine years passed before Hueber got to Australia in 1968. He had planned to collect fossil material in the state of Victoria for research on other Devonian plants. Instead, he took the opportunity to go north to the Fanning River for a few days. In Devonian times the area apparently had been part of a great river delta near the ocean. Trees and other plants had floated downriver, sinking into the delta's sands and muds. Hueber, in the company of Australian geologist Don Wyatt, hastily surveyed some nearby sandstone formations and was encouraged when they found fossil scraps of the genus for which Hueber was searching. Though the material he collected turned out to be poorly preserved, it nevertheless revealed the fact that the plant, instead of being herbaceous of habit, was in truth a tree. But another problem then arose: were the star-shaped strands of fossilized wood part of the tree's roots or part of its branches? Though complex, the anatomy of the trunk did not give clear evidence for orientation of the specimens. Therefore, which way was up?

In 1970 Hueber returned to northern Queensland after his colleague Wyatt had written him that high water in the Fanning River area had cleared a mudstone layer in which two masses of the fossilized Devonian tree were exposed. In Wyatt's opinion, the masses were what remained of the tops of the trees, their branches spreading out through the mudstone. In short order, Hueber collected a considerable amount of the fossil material. But when he got it back to the Museum and examined it, he found it had rotted so extensively before fossilization that it was useless in solving the up-or-down problem.

It was not until Hueber's third trip, in 1975, that he and Wyatt discovered the key 8- by 11-inch chunk of log weathering out of the sandstone ledge. Having studied it closely back at the Smithsonian, Hueber now feels reasonably certain that the anatomy of the specimen is intact and that the orientation of the specimen indicates that the star-shaped strands are the tree's roots.

This discovery helps trace the early evolution of the plant kingdom, for it marks a point in geologic time at which roots can be defined as an integral part of the plant body.

The Magnificent Foragers

Fossil Veins

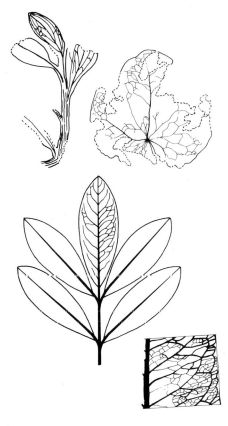

Leo J. Hickey at an early flowering plant site in sediments of the Potomac Formation near Washington, D.C., top. Some of the remains include an early herb that may be the forerunner of grasses, palms, and orchids, shown by drawing, top left; the earliest floating flowering plants, possibly ancestral to water lilies, to its right; and the first pinnately compound leaf and inset of its vein pattern, bottom.

Southeast United States

When flowering plants appeared on the earth over 100 million years ago, many of the older, more primitive plants were forced into extinction. Though this biological alteration marked a critical point in evolutionary history—ultimately making possible man's emergence—surprisingly little data exists on where, why, and how flowering plants evolved.

These questions are precisely those Smithsonian paleobotanist Leo J. Hickey seeks to answer. "Nearly 100 years ago, Charles Darwin referred to the origin of flowering plants as an 'abominable mystery,'" Dr. Hickey relates. "Until recently, paleobotanists were simply matching ancient plant forms with modern ones. But only by studying differences in the shape and form of the leaves and pollen of these plants can we begin to chart changes in plant lineages. I began to use this method with flowering plants after becoming disenchanted with previous 'picture-matching' modes of identification."

Although initially discouraged by the vast number of misidentified fossil leaves described over the years, Hickey felt that their vein patterns might be the key to greater accuracy. He knew that drug companies had developed methods to recognize certain species from the venation patterns of small leaf fragments, and he hoped that this identification process would work for fossils as well.

"Ironically, the Museum sits atop one of the most complete fossil sequences of these earliest flowering plants," he says. "They are contained in an ancient sedimentary deposit called the Potomac Formation, a 15-mile-wide belt stretching from Virginia to Delaware. The Museum's collection of Potomac Formation flowering plants was assembled by William M. Fontaine before the turn of the century. I've been adding to this collection, but today's rapid pace of excavation and highway construction hinders collecting because exposed surfaces are covered up again so quickly."

Hickey, in collaboration with James A. Doyle of the University of Michigan Museum of Paleontology, began to examine both the fossil leaves and pollen of the earliest flowering plants. Their findings indicate that these plants consisted of small shrubs confined to the disturbed areas along watercourses and that their descendants only gradually achieved the stature of trees and spread inland to dominate the earth's vegetation. Hickey's analysis of venation patterns in fossil leaves has clarified trends in plant evolution and provided a timetable for the origin of some of the most important groups of flowering plants in today's world.

The Ice Ages & Before

X. Amateuris populis

They come from all walks of life and, like the scientists who advise and encourage them, they get bitten by the collecting bug. Without the sometimes impassioned energies of these laymen, the Museum's collections—not to mention science itself—would be all the poorer.

Up in Oregon

Northwest Coast

Clayton E. Ray, curator of the Smithsonian's collection of fossil marine mammals—the largest in the world—is lavish in his praise of amateur collectors. "There are so few professional paleontologists and so much ground to cover," he states. "We can seldom be in the right place at the right time when an important new fossil site is uncovered by erosion or a construction project. It's the dedicated amateurs who usually find out about a new locality or spectacular specimen first, and we count on them to notify us so that we can go in and follow up on the discovery."

Dr. Ray's studies currently center on hundreds of fossil sea mammals gathered on Oregon beaches by an amateur collector, Douglas Emlong. Emlong started collecting more than 20 years ago when he was in the 8th grade, after his family moved into a home on the scenic highway that runs along the coast in Oregon's Lincoln County. That coastline is one of the world's richest sources of fossil marine mammals, but prior to Emlong's arrival only a handful of specimens had been collected.

Innate skill and single-minded determination enabled Emlong to put together an imposing personal fossil collection, one that included remains of whales, porpoises, seals, land mammals, turtles, fish, and sea birds. Soon he began writing the Smithsonian and other scientific organizations for information about his specimens, and consulting geologists who could help him decipher the coast's rock formations. The complex geology of the Oregon coast, in which folded and faulted

Amateur fossil collector Douglas Emlong walks along an Oregon beach.

Amateuris populis

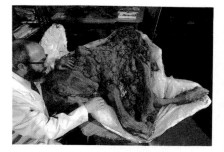

Paleontologist Clayton Ray displays a mummified horse, above. Emlong, right, uses hammer and chisel to get at marine mammal fossil in outcrop.

sedimentary deposits are interbedded with volcanic rocks, is only one factor that makes collecting there difficult. He also had to contend with high cliffs, heavy seas, and incessant rainfall.

In addition, Emlong was handicapped by the fact that he had little money to finance collecting work, which is very often expensive. What money he had he put into his hobby, buying specimens from other collectors in the area, for example, or hiring logging trucks to salvage big blocks of rock off the beach.

"Most of his collecting had to be done in the winter storm season," says Ray, "because that's the time the beaches get scoured clean of sand and good pickings appear on the rocky beach floor. Often Emlong would go out and collect while the storms raged. He's been out on the beach in winds of gale force, knowing that newly uncovered fossils may be buried under sand again before the storm is over. There are many sections of the rock floor along the beach that Emlong has never seen during his 20 years of collecting. But he knows that some day, during some storm, they'll be uncovered."

Emlong transferred his collection to the Smithsonian in 1967 after the Institution reimbursed him with private funds on the basis of the time and expense involved in acquiring, preparing, and housing the collection. After that, he went to work for Ray as a field collector. The more than 1,000 major specimens in his collection, and those he has uncovered since, are now being studied by a dozen scientists at the Smithsonian and at other institutions. "Such a scientific treasure," declares Ray, "will ultimately make an important contribution to our knowledge of the evolution of marine vertebrates."

The Magnificent Foragers

Collecting Insects

National Museum of Natural History

Stinkbugs and flower flies abound. So do *Bombyliidae, Rhinotermitidae,* and *Dolichopsyllidae.* And, of course, *Strongyloptalmyiidae.* These species of flies, termites, and fleas represent but a small part of the Smithsonian's entomological collection, which originated in 1881 with the transfer to the Museum of 50,000 specimens from the Department of Agriculture. Today, it numbers more than 24 million insects and related arthropods.

So large is the collection now that some of the Smithsonian's entomologists no longer go into the field, but devote all their time to deciphering and describing this vast amount of material. Richard C. Froeschner is one such scientist. "I provide evolutionary clues for the field biologist to use in finding how an animal lives, and what role it plays in its environment," he avers.

One of Dr. Froeschner's interests is the study of lace bugs, tiny insects that look like flakes of very fine lace and live upside down, clinging to the underside of plant leaves, from which they suck sap for nourishment. The Museum was given the world's largest collection of these insects—100,000 specimens—in 1957 by entomologist Carl J. Drake, who assembled them while teaching at Iowa State University. Because lace bugs damage a considerable number of cultivated crops, Froeschner is preparing an identification manual on them.

Important collections like the one amassed by Dr. Drake are always sought after by the Museum. "We're constantly on the lookout for good material," says entomologist J. F. Gates Clarke. "Many generous donations are made to the Museum. Occasionally, however, outstanding collections, usually of foreign origin, come to our attention but are available only through purchase. Often we decide to go ahead and buy, particularly if the collection is extensive and possesses significant research value. Of course, it is true that what we spend in this way is but a fraction of what it would cost us to duplicate such collections through field expeditions."

The Museum in considerable measure owes the size (3,500,000 specimens) and excellence of its moth and butterfly collection to outstanding amateurs, one of whom, William H. Barnes, a Decatur, Illinois, physician, employed agents all over the country to collect for him. "He would offer a collector a dollar for every different species of moth or butterfly delivered," recalls Dr. Clarke. "If he was lucky, the collector would return with a great many specimens. Barnes would choose the 100 choicest ones and pay him $100. All told, Barnes spent over $400,000 on his collection. He even had his own private museum and preparator, and was so busy with butterflies that I don't know how he ever took time to practice medicine. When he died in 1930, at the age of 70, he had brought together the finest and most complete collection of North American butterflies ever made—a total of 500,000 specimens."

"Insect collecting intrigues and captivates people, just like art collecting," says Donald R. Davis, head of the Entomology Department. "They become thoroughly dedicated to it as a life-long pursuit. Appreciating that kind of effort and dedication, as well as its practical value, our staff does all that it can to encourage promising amateurs to

A few specimens from the Smithsonian's collection of 7 million beetles, top. Bottom, a Sycamore lace bug, so named because of the damage it has inflicted on American sycamores. Adult lace bugs hibernate under the loose bark of sycamores, and the females lay their eggs on the trees' buds in the spring. Known in America since the 1830s, lace bugs have been plaguing European plane trees, relatives of our sycamores, since the late 1940s.

Amateuris populis

make a significant contribution to science." Citing a relatively unexplored area, Dr. Davis suggests that the small moths of Texas are almost as poorly known as those of South America. "Someone who lives in Texas and is interested in natural history and collecting can, over a relatively short period of time, become an authority on the moths of that area," says Davis. "For example, a man came in a year ago from Louisiana—also a relatively unexplored area for small moths—and showed me moths that we thought lived only farther south in the tropics. I haven't heard from him for awhile, but I certainly hope he's still out there collecting. He taught us something, as I'm sure other amateurs will in the future."

The Smithsonian's collection of 7 million beetles reflects the astounding fact that there are more species of these insects than all the other animals in the world put together.

"We're not amassing all of these beetles for the sheer sake of collecting or pinning them in boxes," says beetle expert Terry Erwin, "although as natural historians we do have an admiration of nature's handiwork. What we are really trying to do is make the beetle collection as complete as possible." According to Dr. Erwin, this requires as many as 30 duplicates of each species so that researchers can perform statistical sampling analyses. The beetle specimens are systematically arranged so they can be efficiently used by all qualified students and scholars, whether they be government agents tracking down a beetle thought to be a potential economic pest, or ecologists compiling background data for an environmental impact statement. "We have a long way to go," admits Erwin. "Some 300,000 beetles have been described to date, and there may be as many as 800,000 species living today."

Paul J. Spangler is the all-time champion water beetle collector, netting more than 300,000 specimens in tropical Asian, African, and South American ponds, lakes, and streams. Dr. Spangler and his wife Phyllis usually get up at dawn and work for 16 hours when they are in the field. At ponds, Spangler wades into the shallows in hip boots and sweeps his net through the aquatic vegetation where the beetles are most abundant. At streams, he stretches a mosquito-netting seine across the water and then kicks over rocks upstream so that the cur-

Three field workers, above, search for aquatic insects on Argentina's Isla de los Estados, an uninhabited island off the tip of Tierra del Fuego. At right, Paul Spangler and a graduate student place a mosquito-netting seine across a stream in rural Maryland.

The Magnificent Foragers

Giant millipedes like the one above are found in Africa and South America. Below, a centipede from the American Southwest devours its prey, which it has already immobilized with venom from its jaws.

rent will sweep dislodged beetles into the netting. He empties the catch onto a sheet, picks up each beetle with a bulb aspirator, and drops it into a bottle of alcohol.

"We get hundreds and hundreds of water beetles every day using these methods," he says. "It's a much different game than 100 years ago, when a collector got down on his hands and knees by the bank of a stream and dipped out beetles with a tea strainer as they swam by, or picked them off of the rocks one at a time with a pair of forceps."

In the field, Spangler also nets material to take back to his fellow aquatic entomologist Oliver S. Flint, Jr., who returns the favor when he is in the field. Dr. Flint is a specialist on caddis flies, mayflies, damselflies, and dragonflies, which—along with water beetles—constitute one of the major forms of life in freshwater ponds and streams. Both he and Spangler are interested in the applications in public health and ecology for the data they assemble on the relationships, origin, and distribution of these little-known insects. Water beetles have been implicated in disease transmission, and changes in the population density of caddis flies and other aquatic insects provide an early warning of water degradation.

"The caddis fly larvae, which attach themselves to boulders and other underwater objects in the stream, have silk glands with which they spin incredibly fine fishing nets," explains Flint. "They use them to strain algae and other microorganisms out of the stream. If the stream begins to show signs of pollution, this algae food source dies off, and caddis fly populations go into decline."

The Smithsonian's myriapodologist, Ralph E. Crabill, Jr., reports that he has a number of amateur collectors who stay in regular touch. "I have one," he says, "who mystifies me. Once a year, as though it were a Christmas present, he—or she—sends me a box of beautifully prepared centipedes collected on California's Monterey Peninsula. Not one word of explanation is ever enclosed, and I've never found out who the person is."

Arachnids (spiders, mites, daddy-longlegs, ticks, and scorpions) and myriapods (centipedes and millipedes) are Dr. Crabill's principal curatorial concerns, his research focusing on centipedes. During his

Amateuris populis

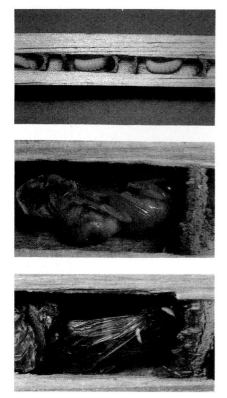

tenure he has taken possession of the R. V. Chamberlin Myriapoda Collection, a gift that at one stroke gave the Museum the largest and most important collection of centipedes and millipedes in the world. Chamberlin, an administrator, teacher, and researcher at the University of Utah for most of his life, was America's most prolific contributor to the study of these animals, accumulating more than 150,000 millipedes and centipedes. As a young man at the turn of the century, according to Crabill, he wandered around Utah's wilds with his collecting tweezers in one hand, and a carbine in the other.

"Chamberlin died unexpectedly in 1967," says Crabill, "leaving his huge collection arranged in a way known only to him. Identifications on the jars are very difficult to decipher. It will be my job for many years to get it all straightened out.

"Why do I study them? They pose tantilizing problems that I enjoy solving," he explains. "Little is known of their ecology or behavior. Their classification is in chaos and always has been. Aspects of their body patterns are little changed by time and provide us with information about arthropod evolution. They're married to the soil and have had to creep to get where they are, which gives us clues to ancient routes of dispersal and continental drift. Beyond that, I like them. They have a certain fearful symmetry."

A solitary wasp's industry and ingenuity in building its nest have fascinated naturalists–both amateur and professional–for centuries. The collections of one amateur observer, George Peckham, form part of the nucleus of the Smithsonian's wasp collection, whose present curator, Karl V. Krombein, has taken Peckham's and others' wasp observations a step further. Dr. Krombein utilizes an ingenious method of documenting the sequence of wasp nest development. He bores holes in the ends of pine sticks, splits them lengthwise, clasps the two halves back together with tape, and hangs them from trees and shrubbery, where they are discovered by wasps searching for cavities in which to prepare their nests. He can then take the sticks apart and photograph at regular intervals. Above, top, full-grown wasp larvae in cells; center, maturing pupa; and bottom, newly emerged adult. At right, Krombein collecting wasps in Sri Lanka.

The Magnificent Foragers

XI. The Practical Naturalist

Increasingly, the doctor of, say, entomology must act like a physician of nature: diagnose and, where necessary, intervene to stem some environmental disequilibrium, or re-establish a lapsed symbiosis, or even stay the course of an epidemic.

Cold Squash Bees

Sacramento Valley; Hawaii

Inert, the squash bees lie in a styrofoam ice chest which sits on the back seat of a car. Captured early in the morning in a Sacramento Valley pumpkin field and tranquilized by the cold, the bees are on their way to the Oakland airport. Before the day's close they will be transported more than two thousand miles and released in Hawaii.

If all goes as planned, the bees will quickly adapt to the islands, building nests and buzzing about to pollinate squash and pumpkin plants. Although many insects and bees visit the flowers of these plants for their nectar, most of them do so on a casual hit-or-miss basis. Only the squash bee can be counted on to make regular trips from flower to flower, effectively and efficiently cross-pollinating.

"Squash bees have this special talent," says Smithsonian entomologist Paul D. Hurd, "because they evolved in close mutual association on this continent with pumpkins and squashes (*Cucurbita* plants), becoming specially adapted to pollinate them."

Until recently, both entomologists and growers were unaware how essential these bees were to *Cucurbita* pollination. Attempts had been made to introduce native U.S. pumpkin and squash plants to other areas of the world in the hope that honeybees or native wild bees would pollinate them, or that man himself could do it by hand. It didn't work well. Lacking resident squash bees, the plants produced little. Large-scale cultivation of these important food plants outside the Western Hemisphere seemed out of the question.

"When my colleagues and I made the discovery that these bees were the only effective pollinators of squash plants," says Dr. Hurd, "we decided to try to restore the squash-bee partnership in the Hawaiian islands. There are squashes there but no squash bees, and the yield is almost one-half what it is in the Americas. If the experiment is successful, we believe that it may be possible to repeat it, doubling the production of these vital food sources elsewhere in the world."

Fellow entomologists E. Gorton Linsley and A. E. Michelbacher assisted Hurd in deciding which species of squash bee to export to Hawaii. After studying the distribution, ecology, and behavior of 21 kinds of squash bees that range through North and South America, they selected *P. pruinosa*. It proved to be an efficient pollinator of almost all domestic *Cucurbita* plants as well as possessing the ability to survive in a wide variety of climatic and topographical conditions.

"On the plane flight to Hawaii," Hurd recalls, "Dr. Michelbacher kept wanting to open up the ice chest to check on the condition of the captive bees. I warned against it because I was afraid that some of the bees might escape and start buzzing the other passengers. It could have happened, because when we got to Hawaii, an Agriculture Department official asked that we open the box for the customary inspection. When we did, out flew 15 or 20 bees. Fortunately, the rest of the bees stayed in the box. By late afternoon we had released hundreds of bees at various locations on Oahu and on the neighboring island of Hawaii. We're now monitoring these sites to see if breeding populations of squash bees will succeed in permanently establishing themselves. If they do, the productivity of squash plants will increase nicely and we will have a precedent for trying transplants elsewhere."

A cluster of six transplanted male pruinosa *bees visits a male squash flower in Hawaii, eagerly seeking nectar after a long airplane flight from California. Selected for squash pollination experiments, they were collected in the Sacramento Valley, where they live near pumpkin and squash fields. Hawaii has no indigenous squash bees and, as a result, its squash plants produce little.*

Mayans and Lichens

Central America

Deep in the interior of Yucatán, Guatemala, and Honduras lie the monumental remains of the Mayan civilization, which mysteriously collapsed in 900 A.D. Jungle growth quickly swallowed up the large stone tablets set up at the bases of ceremonial pyramids. Now brought to light again, the monuments' complex inscriptions are being deciphered, providing dates and notations of astronomical, historical, and religious events recorded by the Mayans.

Archeologists in Central America recently noticed, however, to their alarm, that dense infestations of lichens were eroding and blurring the faces of the monuments. Smithsonian botanist Mason E. Hale, Jr., an authority on tropical lichens, was quickly called in to study the problem at Quiriguá, Guatemala, and Copán, Honduras.

"I found the lichen growth posed a two-fold threat," he says. "Moisture in the lichen cover was breaking up the rock crystals, and lichen acid excretions were disintegrating the rock minerals. As long as the monuments were protected by the shade of the lush jungle canopy and undergrowth, or covered with vegetation and dirt, the lichens did not grow on them. But when archeologists cut the dense forest away to expose the monuments to the open air and sunlight, conditions became favorable for lichen growth."

Asked to come up with a method for controlling infestation, Dr. Hale experimented with mild bleach, borates, and phenolic solutions that have been used in England and France in recent years to combat lichen growth on tombstones and buildings. Walking from monument to monument through the acres of ruins, he applied a sodium hypochlorite commercial bleach that he carried in a shoulder-pack garden sprayer. The more sensitive lichens died almost immediately and within four months could be removed from the rock surface with a soft brush. More than one application was necessary for some of the tougher lichens, but complete cleaning of the monuments has now been accomplished. Currently, Hale is looking for another spray solution that will leave an active residue and prevent the airborne lichens from colonizing the monuments a second time.

"Only recently have we awakened to the fact that the black appearance of old marble buildings and monuments in Europe and America cannot always be blamed on particulate pollution," says Hale. "If you think that by cleaning up air pollution you can stop this blackening, you'll be fooled. The buildings will still turn black. The real culprits are microscopic airborne crustose lichens."

Before and after: profuse lichens cover a Mayan monument, top, at Quiriguá, Guatemala, while the stela pictured below it, also at Quiriguá, has been cleaned and brushed. Bottom right, a close-up of one of the lichens that have been plaguing monuments in this area.

Leaf Miners

The tracks on these tropical leaves represent trails left by only three of the thousands of unnamed families of South American leaf miners.

Virginia's Great Dismal Swamp

Donald R. Davis has dozens of covered jars in his laboratory that are full of blotched and discolored leaves bearing bent, serpentine, and star-shaped tunnel markings. The leaves he collected in Virginia's Great Dismal Swamp. The marks are the work of leaf miners, tiny moth larvae that tunnel between the upper and lower leaf surfaces of a plant or tree. Miners cause so much damage that biological controls are now being sought to control them.

As a prelude to research, certain basic facts about these insects must be established. Miners are so tiny and difficult to observe that it is not known how to identify many of them; not much information is available on their behavior, or what plants various miner species favor. Dr. Davis is finding out about behavior and food sources by rearing miners in his lab, watching as they change from egg to larva to pupa, and finally to adult moth. Figuring out the proper identification is another problem, a crucial one.

Existing names and descriptions of leaf-mining moths are riddled with errors because they date to the 19th century, a time when investigators classified moths or butterflies according to the color pattern of the wing. Nowadays, the insect's skeleton is seen as a far more reliable part of the body on which to base classification. In this light, Davis is re-examining the anatomical details of miner types. When finished, he plans to publish an identification guide to illustrate and describe all known moths in the four principal North American leaf-mining moth families, placing them in their correct evolutionary relationship.

"A lot of people wonder why we spend our time here describing insects," Davis said recently. "What they don't realize is that food production is basic to our existence—and insects are our chief competitors for that food. Every year they chew their way through hundreds of millions of dollars worth of field crops, orchards, and food in storage —not to mention rugs, furs, clothing, and other animal products.

"We enter the picture because no one who finds an insect eating crops in his fields can even begin to do anything about it until he finds out what it is. That's what much of our work is all about. It's extremely practical, actually. It enables a scientist to examine an insect and to tell what it is, where it is likely to be found, what it feeds on and, perhaps eventually, who its natural enemies are."

State agricultural officials will often turn to the Smithsonian collections for help in identifying moths. Not long ago, they sent to the Institution a batch of enigmatic moth specimens captured near Cleveland, Ohio. The insects were tiny and slender, with a light yellow band around their abdomens and tufts of bushy, hair-like scales rising from the tops of their heads. It took long searching through the literature and collections before Davis discovered what they were: general stem moths of a variety which had heavily damaged winter wheat and rye crops in Russia over the past 20 years. The moths, which have since turned up in two other states, probably entered the country via shipping through the St. Lawrence Seaway. Determining the moth's geographic distribution in Europe and Asia, Davis swiftly warned agricultural agencies that if methods of control were not found, it could become a serious pest in major grain-growing areas of North America.

Trouble in the Chesapeake

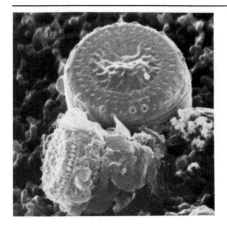

The aggregate cluster above, composed of mineral particles and organic matter with attached diatoms, is a prime suspect in Jack Pierce's investigation of estuarine plant life killers. Aggregates like this are common in the sediment of the Rhode River, which flows into Chesapeake Bay. These microscopic aggregates (the largest particle in the cluster is 5 microns, or 5 millionths of a meter across) contain unusually high concentrations of metals and herbicides.

Annapolis, Maryland

The Smithsonian's J. W. Pierce is concerned. Underwater vegetation in the Chesapeake Bay, once so lush and diverse that it was a problem to control, is dying off. Canvasback ducks, Canada geese, and other migratory waterfowl that depend upon it as forage are leaving the area. Essential nursery grounds for finfish and shellfish are foundering at the same time.

Dr. Pierce has been working on several theories as to why the aquatic vegetation is disappearing. Having all but ruled out disease as the killer, he sees two remaining possibilities: turbidity, which is associated either with land sediments in the water or with increasing biological productivity that cuts down light penetration; and herbicide run-offs from surrounding agricultural lands. Pierce, in collaboration with two scientists at the Chesapeake Bay Center for Environmental Studies, has designed an experiment to determine how extensively the Bay is contaminated by some of these toxic chemicals.

At the Center, a 2,400-acre field research station located on the Rhode River seven miles south of Annapolis, Maryland, and maintained by the Smithsonian and a consortium of universities, Pierce and other investigators built weirs along a dozen small streams emptying into the estuary. The weirs divert a proportionate amount of every week's downstream flow into large bottles. The water samples and their suspended sediments—along with stream-bottom sediments—are then analyzed for organic and mineral matter, selected herbicides, and trace metals. Periodically, water and sediment samples from different sites in the estuary are also analyzed.

Examining the sample material under a scanning electron microscope, Pierce discovered the presence of sediment aggregates composed of fine mineral grains and organic matter such as phytoplankton and bacteria. These aggregates absorb some trace metals and herbicides, creating microenvironments that bear little resemblance to the environment of the stream or the estuary as a whole.

"Dissolved herbicide concentrations are low in the water itself," Pierce explains. "But in the sediment aggregates, we found that herbicide concentration is greatly enhanced—as much as 110- to one million-fold. Evidence exists that these aggregates, with their high herbicide content, are settling on the leaves of the aquatic plants as well as in the bottom sediments around their roots, possibly causing the die-off."

According to Pierce, the sooner a technique to control the transport and deposition of the aggregates is developed, the sooner some of the major pollutants can be eliminated.

XII.

Meanwhile, Back at the Museum

. . . it all starts making sense. The collections are the touchstones of nature, be they fish or fowl.

Invertebrates

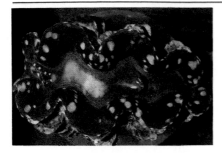

Giant Indo-Pacific clams such as the one above were collected by Joseph Rosewater at Eniwetok Atoll, Marshall Islands. The largest on record came from Sumatra, measured 4 feet 6 inches in length, and weighed 507 pounds. Few documented reports exist of a giant clam killing anyone, but the huge jaws can close swiftly if the animal is disturbed. The Georgia crayfish, opposite, is a species that lives in shallow stream rapids.

Department of Invertebrate Zoology

The Smithsonian's immense ocean life reference center began in 1857, when the newly chartered Institution received for safekeeping the bottled and kegged specimens gathered by the ships of the celebrated, around-the-world U.S. Exploring Expedition. The Museum's collection has been growing ever since, and now includes sponges, sea urchins, crabs, sea fans, crayfish, coral, worms, ostracodes, amphipods, and a multitude of other invertebrate animals.

"Never in a million years could we publish enough words or pictures to fully understand these organisms," says J. Laurens Barnard, curator, Department of Invertebrate Zoology. "That's one reason they must be retained. Also, if we don't keep them, future generations will not have material to study that can give them a chronological perspective on the impact man has had on our waters. Only now are we beginning to face the fact that man is having an adverse effect on more than just birds and mammals—many of the inconspicuous animals of the sea are also headed for extinction."

Dr. Barnard's own work focuses on amphipods, tiny shrimp-like animals that abound in the world's waters and serve as an important source of food for fish and other aquatic life. Amphipods have taken on a new significance because of their crucial role in monitoring programs that guard against pollution off various coasts. Barnard's studies of these animals have made them so biologically well known that scientists can determine their levels of tolerance to oil and sewage. Rapid die-offs or population changes among certain amphipods are usually a warning sign that the marine environment is deteriorating.

"Our aquatic collection is expanding faster than others in the Museum—with the possible exception of insect specimens—because of the great diversity of animal species in the oceans and freshwater bodies of the world," says Barnard. "Acquiring this material is extremely expensive since collecting often requires the use of seagoing ships, deep-sea dredges, and submersibles."

Shellfish make up the invertebrate collection's largest single unit. There are more than 12 million mollusks, many gathered by the Smithsonian's Joseph Rosewater and Harald A. Rehder. Dr. Rehder estimates that on field trips to the South Pacific he has dredged and gathered—on reefs, in tidal pools, and in the ocean itself—at least a half million mollusks. This collection, he says, uniquely defines the distribution and distinct character of all Pacific shell life.

Meredith Jones and his colleague, Marian H. Pettibone, specialize in polychaete worms, tiny animals of the seas that, like amphipods, are seen by scientists as sensitive indicators of environmental change. Smithsonian collections hold hundreds of thousands of the creatures.

"They have a simple body plan that can be divided into 65 basic categories or families," says Dr. Jones. "Each of these can be further subdivided, a fantastic amplification of diversity on a central theme. We must learn to differentiate between all these animals, for the basis of all experimental ecological work is in knowing what species you're dealing with. If an ecologist is trying to cope with pollution, and 90 percent of the animals he observes are unknown, he's up the creek."

The Smithsonian's collection of crustaceans—crabs, shrimp,

Meanwhile, Back at the Museum

Jars of crustaceans in the Smithsonian collections, above, gathered by 19th-century exploring expeditions. A Panamanian worker, opposite top, cleans a sieve used by Meredith Jones to sample worms and other invertebrate animals from the mud flats on the Pacific side of the Isthmus of Panama. Below, Colby A. Child collects mollusks, crustaceans, and other invertebrates from tidal pools on the Atlantic side of the Isthmus.

lobsters, and such—is unsurpassed. More than five miles of shelving is required to hold it all. "I can't begin to estimate how many different kinds of crustaceans we have here," says the Smithsonian's Raymond B. Manning. "The collection is like a library—one of the finest in the world. The specimens are the library's books, and the more they are studied the more valuable they become."

Field work by the staff is continually adding to the collection. Ascension Island in the South Atlantic swarms with both land and shore crabs. On two trips there, Dr. Manning rounded up a remarkable variety of these clawed animals. Because of Ascension's isolation and relative geological youth—estimates make its age no older than one million years—it is an excellent natural laboratory. It is possible to study where its crabs and other marine animals originated, how this life was dispersed and carried to Ascension by currents and other means, and how the animals have adapted to their environment since their arrival. Manning and Fenner A. Chace study the island's crustaceans, Dr. Rosewater the mollusks, and David L. Pawson the echinoderms.

Dr. Pawson estimates that approximately one new animal species has colonized the island every 10,000 years, and speculates that these organisms arrived in a number of ways: borne by ocean currents, carried on floating objects, blown in on the wind, or even transported by birds. Man is also implicated. During World War II, Ascension became a refueling stop for transoceanic flights, and more recently it has served as a missile tracking station. As a result, humans may have inadvertently transferred numerous forms of life to the island.

The Museum has the largest corps of crustacean experts in any scientific institution in the United States. Through most of this century the patriarch of this group was Waldo L. Schmitt, who first came to the Smithsonian when the Institution opened its present Natural History Museum building in 1910. Continuing his indefatigable travels

Meanwhile, Back at the Museum

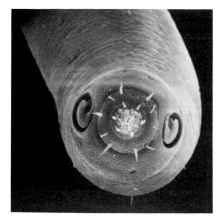

A scanning electron micrograph of the head of a deep-sea nematode that measures less than a millimeter in length. Invertebrate zoologist Duane Hope studies the systematics, evolution, and comparative and functional anatomy of marine nematodes, microscopic worms that take myriad forms and live in watery habitats ranging from pockets of melted water on glaciers to hot springs, lakes, estuaries of major rivers, and oceans. The closed oral opening of this specimen is surrounded by three circles of setae, believed to be sensitive to touch. Nerve endings at one end of each of the spiral grooves on both sides of the neck may be sensitive to dissolved chemicals in the animal's aquatic surroundings.

after his retirement in 1957, Dr. Schmitt made his last major collecting trip in 1962, bringing back 25,000 marine specimens from the waters off Antarctica. The Smithsonian's crustacean collections by then were becoming the largest in the world, and the staff was growing. By the mid-1960s, in addition to Schmitt, Manning, Chace, and Horton H. Hobbs, Jr.—authorities on decapods, the order that includes familiar restaurant crustacea and crayfish—other scientists working on smaller crustacean orders had also joined the staff. Thomas E. Bowman, for example, studies copepods and isopods, and Louis S. Kornicker specializes in ostracodes. The midget crustaceans these men study are much more abundant than the decapods, and descriptive knowledge and nomenclature for these animals are urgently needed. Other scientists snap up the basic taxonomic work of these men as soon as it is published, applying it to practical problems.

Dr. Hobbs has been collecting crayfish for taxonomic studies since his college days in the early 1930s. These freshwater creatures occur in the greatest abundance and diversity in the United States in the southern Appalachians, where they are most easily caught at breeding time in the early spring. At this time of year Hobbs is often found in Tennessee or Arkansas down on his knees along a creek, digging away at a hole with his hands until he gets down deep enough to thrust in his arm and pluck out a crayfish. They resemble lobsters, but are only 3 or 4 inches long. Some species of crayfish live exclusively in caves, where they have developed delicate body forms and become albinos. With his son, also a crayfish expert, Hobbs has explored dozens of caves. The crayfish he captures he brings back to the Smithsonian so he can observe their breeding habits. "They get along well on a diet of fish-flavored cat food," he reports.

Another Smithsonian crustacean expert, Roger Cressey, studies the taxonomy of parasitic copepods, commonly referred to as "fish lice." Most of the nearly 2,000 known species of these fish parasites live in the oceans. Dr. Cressey works with copepods that live on many varieties of marine fish, but he is especially interested in those found on sharks. Not long ago, Cressey cruised the Indian Ocean in a research vessel to collect copepods, which meant that first he had to catch sharks. Early every morning a cable 8 miles long was strung out into the ocean, buoyed by hundreds of hollow rubber balls. More than 500 baited hooks dangled 50 feet deep into the ocean from the cable.

"The trot lines were hauled in every afternoon," comments Cressey. "This was very hard work, considering the 1,000 pounds of dead weight you might have at the other end if you had hooked a big one. In six weeks we caught about six dozen sharks. Many of them were pulled up alive and then killed by a blow on the head with a wooden mallet. We had to be very careful, for even after a shark is technically dead it can snap its jaws and thrash convulsively when touched. Many people have been bitten that way. I did not reach into 'dead' sharks' mouths to search for copepods until I had propped open their jaws with a stick. The only time I was attacked was when a dead shark I was working on whacked me in the legs with its tail."

Cressey left the ship after six weeks and went to Madagascar, where he spent two months doing parasite research at a French oceanographic laboratory. In addition to his own collecting, he paid the natives to bring him the sharks they had caught. The natives

The head of a sand shark, top. Center, a cluster of parasitic copepods near the pelvic fins of a shark. Fishermen hauling in a trot line in the Indian Ocean, bottom.

hauled their catch back to him in their outriggers, some of the sharks still alive and thrashing, and as large as the canoes.

On other occasions Cressey collected sharks off Sarasota, Florida. "We hung the trot lines on a cable suspended between two 50-gallon oil drums," he explained. "During the time I was there we collected 200 to 300 sharks representing 16 different varieties. The largest were hammerheads and tiger sharks, both kinds measuring up to 12 feet in length. To catch them we used huge hooks a foot long made of half-inch-thick steel. Sometimes we'd pull in the lines and find that sharks had completely straightened out the hooks and gotten away. It's difficult to imagine sharks that strong, but the big ones are."

Cressey is making a world-wide survey of the evolutionary relationships of copepods to sharks. He knows that for every different species of shark he catches he will find a different copepod fauna. "What happened is that different kinds of sharks evolved over millions of years, changing in form as they moved into new underwater environmental niches. The copepods that lived on these sharks moved with them and also changed, although not as fast as the sharks did. Thus, when we find two related copepods on two different kinds of sharks, it is supporting evidence that there may be a close evolutionary relationship between the two sharks even though they now bear little resemblance to one another."

Fishes

Stanley Weitzman observes South American flying characins in an aquarium, above.

Department of Vertebrate Zoology

Adding to the bounty of the U.S. Exploring Expedition, U.S. Fish Commission ships plying American coastal waters in the late 19th century collected and sent back to Washington tons of fish specimens. This gave the Smithsonian one of the largest study collections of fish in the world. Today, the inflow of specimens continues unabated, largely through exchanges with other fish research centers and through the field activities of Smithsonian scientists. To wit:

Thousands of fish have come to the Museum as a result of a research project headed by ichthyologist Robert H. Gibbs, who is trying to find out how fish populations react to industrial chemical wastes dumped in the Atlantic Ocean. The Gibbs team caught fish at different levels at a dump site abutting the continental slope, 100 miles off the coast of New Jersey. As the dumping continues, the data amassed will provide a baseline to make it possible to detect resultant changes, if any, in the fish and animal populations at the site.

The streams and lakes of South America hold hundreds of undiscovered fish species. Stanley H. Weitzman investigates the biogeographic and systematic relationships of these fishes so that the aquatic resources of the continent can be better understood. A tank full of small Peruvian flying fish is the object of one of his studies: he is observing how they move their pectoral fins and launch themselves out of the water to evade pursuers.

Victor G. Springer, on the other hand, specializes in studies of the fishes of the tropical Indo-West Pacific, often returning from field cruises with dozens of undescribed species. It sometimes happens that his laboratory studies pose challenges that propel him right back into the field. On one occasion he returned from Australia's Great Barrier Reef with a batch of yellow, white, and black "blenny" fish that he assumed were all the same. But when he examined them closely, he realized that he had confused two different species. One appeared to be copying the coloration and markings of the other.

Because mimicry had rarely been observed in fish, and since Dr. Springer wanted to learn how it works among blennies, he made dives at a Gulf of Aqaba coral reef. There, at depths of 10-30 feet, he found a variety of blue, black, and yellow blennies that resembled each other closely, and lived in association with one another, but that had different modes of behavior. Some swam or hovered over a territorial home site on sandy flats between coral outcrops. Others spent a large part of their time perching motionless on rocks and dead coral, swimming out occasionally to forage for something to eat and then returning to the same spot—darting into holes if threatened. A third blenny type swam around slowly, occasionally rushing at Springer, trying to bite him.

After hours of observation, Springer figured out what each look-alike blenny was doing, and why: "Blenny number one, hovering over a home site, is a fang-toothed blenny. It is protected by a pair of venomous fangs in the lower jaw (no other fishes are known to possess such fangs), and by its bite is able to force predators that have engulfed it to immediately spit it out, unharmed. A predator that has had this experience tends not to bother a fang-toothed blenny again.

According to Victor Springer, the female "wolf-in-sheep's clothing" blenny, top, owes its success as a predator to its resemblance both to the fang-toothed blenny, below left, and the "helpless mimic," below right.

Nor is the same predator likely to pick on fish number two, the comb-toothed blenny. This blenny has no defensive fangs, but relies for protection on the holes it hides in, as well as on its ability to mimic the fang-toothed blenny. The third blenny in this mimetic complex is the snake blenny, a predator. Because it resembles both the fang- and comb-toothed blennies, the other fish on the reef perceive it as a fish that will not attack. Moving in close to the other fish with its 'wolf-in-sheep's-clothing' disguise, the snake blenny takes a bite out of them," reports Springer. "As it tried to do to me a few times."

Meanwhile, Back at the Museum

The National Herbarium

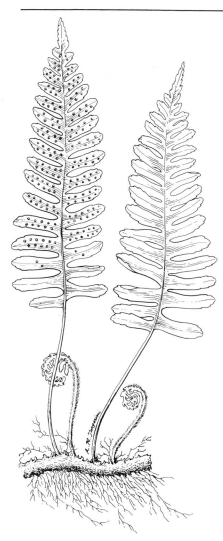

Drawing of Common Polypody, a fern found in damp wooded areas of eastern North America. Two adult fronds are shown with fruit dots visible on the left frond.

Department of Botany

Millions of flowering plant specimens and hundreds of thousands of ferns, mosses, lichens, and fungi are housed in tall metal cabinets at the Smithsonian. As new dried plant material arrives at the National Herbarium it is sent to a sealed room and fumigated in its wrappings to kill any insects that might be present. Then the plants are turned over to technicians who tape and glue specimens onto sheets of rag paper, attach typed identification labels, place the sheets in manila folders, and store each folder in a cabinet arranged according to the plant's family and the area of the world in which it was collected.

Government agencies, universities, and other herbariums throughout the world send material to the National Herbarium to be identified by having it matched with specimens already on file in Washington. "Curators here routinely identify as many as 3,000 plants a year," says fern specialist David Lellinger. "It takes a third of our time, but we're happy to do it. In return for making the identification, we usually get a duplicate of the plant, thus improving our collections. While doing these identifications we sometimes run across new or clarifying material that aids our research, as well as that of others. For example, I've been working on plants acquired by the National Institutes of Health from a collector in the tropics. Their tests show that the plants may have potential as an anti-cancer drug. Now NIH wants more of this material, but before they can work with it, they must have the plant positively identified. That's where we come in."

"Because of our huge herbarium collection," adds Dan H. Nicolson, a specialist on the flora of Nepal, southern India, and Dominica, "we're uniquely equipped to identify and classify plants. Classifying a plant means fitting it systematically into one family or another, and to do this you have to determine the characteristic parts of the plant's structure that distinguish it from other plants."

Plant anatomist Richard Eyde attempts to deal with such difficulties by microscopically examining flowers and fruits that have been rendered transparent either by exquisitely thin slicing or by chemical treatment. His investigation of flowering plants, such as dogwoods, aims to answer such questions as these: which species retain the most primitive flower structure? What other plants are most similar in microscopic features of flower and fruit? Do these similarities seem to reflect common ancestry when all other traits, such as chemical makeup and pollen characters, are taken into consideration?

By far the smallest botanical specimens in the collections are pollen grains. Joan Nowicke's research is devoted to the study of these minute objects which carry a flowering plant's male sex cells. She takes pollen samples from the plant collections in the Herbarium and prepares them for examination with a light microscope by making permanent glass slides. Most pollen samples are also examined and photographed at very high magnification (enlarged 5,000 to 15,000 times) with scanning electron microscopes, which yield photographs that show the intricate surface of the grains in great detail. In many cases these photographs serve as fingerprints for identification and classification.

One small section of the Smithsonian's Herbarium is devoted to

The Magnificent Foragers

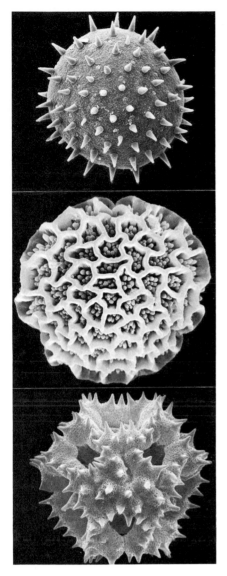

These fine pollen grains, top to bottom, were taken from hibiscus, bougainvillea, and dandelion, respectively. Pollen grains have an elaborate and detailed structure, and their variation makes them useful in classifying plants.

plant specimens of the Washington, D.C. area, an exhaustive historical reference to the capital's richly varied flora. Field work in the Potomac River Valley is carried on by Stanwyn G. Shetler, who is monitoring the disappearance of unique wildflowers such as the pitcher plant, whose boggy habitats are vanishing around major cities.

Among the botanists at the Smithsonian, many have worked extensively on South American plants. Richard Cowan studies legumes (the bean family), which in South America include many excellent woods that can be selectively harvested; Harold Robinson specializes in mosses and sunflowers; and Laurence Skog works with African violets, a flower that occurs in almost 2,000 varieties in the tropics. Joseph Kirkbride, meanwhile, is engaged in pioneering studies of *Rubiaceae*, a plant family whose many varieties—including the coffee plant—make up a major part of the ecological composition of the equatorial regions.

Smithsonian botanists continue to be drawn by the enormous diversity of South American plants, and the fact that some of the oldest and most primitive forms of vegetation on earth flourish there.

John Wurdack, a specialist on the *Melastomes,* one of the larger families of flowering plants, works primarily on plant material from South America. "I can't possibly keep up with all the new species being found," he admits. "It's true that there is a lot being lost through development, but at the same time there are still unspoiled regions that botanists have never seen before—Brazil's northwestern Matto Grosso area, some of the mountains of Ecuador, much of Venezuela's 'lost world' area, and many parts of Peru and Bolivia. These are all places I'd love to go to. In a way, they are botanical paradises."

The lush Andean lowlands of Peru and Brazil attract Dieter Wasshausen, whose specialty is *Acanthaceae*, another large family of tropical flowering plants. "The purpose of my explorations is not only to make botanical collections but to observe and record pollinators associated with this group," Dr. Wasshausen says. "These observations in turn will help to explain the different breeding systems used for maintaining the species."

South America is also a great center for bromeliads, the family that includes pineapples. Botanist Lyman Smith specializes in bromeliads and has prepared a classic three-volume systematic study of these plants. Not long ago, Dr. Smith stepped in as a consultant during an outbreak of malaria in southern Brazil. "Poor ground drainage was blamed until it became apparent that the mosquitoes were breeding not on the ground, but in bromeliads," he explains. "Anchored on treetop branches, the leaf basins of these 'air plants' hold as much as a gallon of water, and animals ranging from protozoa to frogs inhabit those miniature ponds. Examining the 90 varieties of bromeliads growing in the area, we located a handful of species that were significant mosquito hosts and taught the Brazilians how to identify them. With this information they were able to concentrate their spraying on problem plants."

Birds

A stork, above, feeds in a Javanese pond. Right, some of the bird collection's smallest eggs–those of hummingbirds–are compared with larger, brightly colored tinamou eggs.

Department of Vertebrate Zoology

Almost from its beginnings, the Smithsonian has been a world center for the study of birds. A long roster of distinguished ornithologists, including Spencer Baird, Elliott Coues, Robert Ridgway, Alexander Wetmore, and S. Dillon Ripley, contributed to the growth of its reputation. Collections made during boundary and railroad surveys in the 19th century, and those of the Biological Survey and the Fish and Wildlife Service, have made the Institution's holdings of North American birds the largest in existence. There are also outstanding collections from the West Indies, Central America, Southeast Asia, and other parts of the world.

No matter how old, the specimens continue to be studied by new techniques and from fresh viewpoints. The emphasis today is on studying the systematic relationships of major groups of birds and various aspects of their biology. Using the specimens, researchers can study population samples to determine how new species of birds evolve; correlate anatomical variations with behavior and environment to explain the evolution of adaptations; or analyze feather molt and learn how it relates to migration and the breeding schedules. In both a general and a specific way, the collection makes it possible for researchers to make inferences about the ecological relations of birds. To do all this requires a series of specimens of each bird species—not just one or two individuals—because of the variability within each species.

"In addition to research," says the Museum's Richard Zusi, "the collections are used to provide identifications relating to law enforcement, aircraft collisions with birds, archeology, fossils, and medical research. They're also useful because they document the history of environmental and faunal changes. For example, our eggshell collection is consulted by scientists who want to see how much shell thickness has changed since the introduction of DDT into the environment. A collection shouldn't become static," he insists. "If it doesn't grow, we can't continue to make historical comparisons."

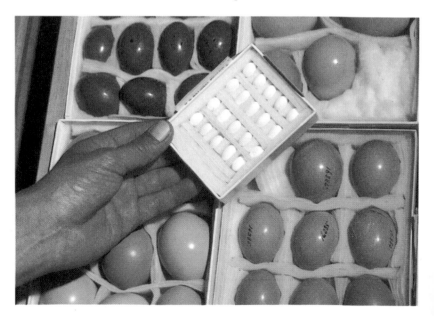

Reptiles and Amphibians

Department of Vertebrate Zoology

George Zug reaches into an aquarium and removes a small frog. Dipping the frog's hindquarters into a dish of ink, he places it on brown wrapping paper spread out on the floor. Abruptly, the frog leaps away, leaving a blotch of ink on the paper each time it lands. When the frog completes a series of five jumps, Dr. Zug returns it to the aquarium, then measures the distance between each ink blotch.

A herpetologist in the Museum's Division of Reptiles and Amphibians, Zug conducts research on how far frogs leap relative to their body proportions. "The good jumpers—like the bullfrogs—have long hind legs and can leap 10 to 12 times their body length; whereas the poor jumpers—like toads—have short hind limbs and can only leap two or three times their body length," says Zug. He has tested more than a hundred different kinds of frogs, and at the same time he has documented the leaping with high-speed cinematography, a technique that makes it possible to observe in slow motion how the frogs use their muscles when they spring.

The frogs used in the tests are then skeletonized and added to the division's collection, which now numbers nearly 300,000 salamanders, frogs, turtles, crocodiles, snakes, lizards, and other less familiar reptiles and amphibians. The majority of these specimens are preserved whole and stored in jars of alcohol in the division's collection room. Cabinets in the same area contain skeletal material like turtle shells and skins of large snakes, lizards, and crocodiles.

"We're still growing," says Zug, who curates the collection along with a colleague, Ronald Heyer, "though we've had to become quite selective. Rather than accepting material from amateur naturalists who collect and preserve animals for enjoyment only, we're trying to act as a depository where scientists can place the materials of their research. This can be anything from a single frog containing a new parasite up to a massive collection of 100,000 salamanders.

"Once in our collection, the frog and salamander material can be used to check the accuracy of the scientist's work. It's also useful because if someone wants to do an ecological study of salamanders he won't have to spend half his life collecting the vast numbers of salamanders needed. We'll already have them here. Don't forget, too, that 100 years from now, if someone wants to study that collection, they'll have a record of what salamanders looked like in the 20th century. They can use it to see if any changes have taken place over the years in size, sex ratio, food habits, or geographical distribution."

A jumping frog such as the one above can move from a rest position to aerial suspension in eight-tenths of a second. It streamlines its body by depressing it, then jumps at an almost 45-degree angle to maximize its jumping distance.

Thieves and Ethics

Department of Anthropology

During the exploration of Latin America in the late 19th and early 20th centuries, thousands of scientifically important antiquities came to the Smithsonian. Over the past 25 years, the Institution has continued to legally acquire objects from Mexico and Central and South America. But, at the same time, it has refused to bid with other museums for the flood of pre-Columbian art material plundered from ancient sites and smuggled illegally into the United States for sale.

Clifford Evans, the Smithsonian's specialist on Latin American archeology, has championed the Institution's stand against trafficking in illegally excavated and exported archeological antiquities. Because the Smithsonian was founded as a research institution and not as an art museum, asserts Dr. Evans, archeological collections have had a different history here than at many other museums. "In our early days we had very little money for purchases," he says. "In the long run our poverty proved valuable to us. We never fell into the habit of purchasing undocumented archeological materials. Some individual objects came to us as donations, but the majority of our material was acquired by expeditions and was well documented scientifically."

When Evans came to the Smithsonian's National Museum of Natural History in 1951 and began to think of modernizing the Latin American exhibits, he saw a number of gaps in the collections that required filling. When it became evident that he needed some material from the early Peruvian Chavin culture, it was suggested that he try to raise funds to buy it. A fine pot from that period sold then for $1,000. Today it would cost $10,000.

"My feeling then, as now, was that it is foolish to spend such sums for unscientifically collected objects. It makes more sense to take a photograph of the pot or have a replica cast, and save the money for a scientific expedition.

"This not only saves expenses; it also keeps us from competing with other museums, driving up the prices of smuggled objects even further. It's the demand that creates the market. The boom for pre-Columbian 'primitive' art began to spring up all over the world after World War II. Lots of people had money, and they saw the purchase of this material as an excellent investment. Many museums bought and built up elaborate collections of pottery, figurines, and stone sculpture—but they were collections without a shred of data.

"As an archeologist it is the data that I value. I don't look at an object to admire its beauty. What interests me is what the object can tell us about a society and how its people lived. If it is to reveal this story it must be excavated scientifically in context with thousands of other bits and pieces of information.

"By digging up these pre-Columbian pieces and selling them, the looters are destroying scientific data and with it the history of the peoples of these countries. The roots of the Mexican, Guatemalan, Ecuadorian, Bolivian, Peruvian, and north Chilean cultures are Indian. Large populations of Indian people still live in these lands today. There is a continuity with the past. So when you talk about protecting an archeological site, you're talking about protecting the roots of the present-day population."

Beginning in the 1920s, Mexico and the Central and South

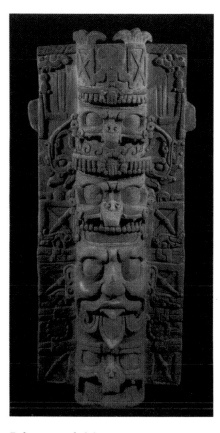

Palenque-style Mayan pottery incense burner from the Netzahualcoyotl region of Chiapas, Mexico. Five similar specimens were confiscated in 1967 by the Instituto Nacional de Antropologia e Historia from persons who were going to ship them illegally out of Mexico to the international art market. This one was presented to the Smithsonian that same year by the Mexican government in appreciation of the fact that in 1908 the Smithsonian returned to Mexico a rare carved hieroglyphic stone from Palenque.

American countries began to pass laws to protect their national patrimony. The legal basis of patrimony in these countries is derived from ancient Spanish law that holds that everything below the ground on a property belongs to the crown. This Spanish concept of law gives a legality to the protection of archeological sites that cannot be attained in the United States, where laws have English antecedents.

The policy at the Smithsonian has been to respect the laws of these countries and not to get involved in buying, collecting, or exporting objects, or in looking the other way. "Defending these laws," Evans notes, "has made us unpopular with collectors and art museums. But by refusing to accept objects that do not have a legal origin, the Smithsonian has earned the good will of Latin American countries. In return—at a time when American archeologists are being barred from working in these countries because of antiquity law abuses—the Latin Americans have allowed Smithsonian scientists to enter their countries and do research. In some cases they have also allowed the Institution to keep scientific collections. When my wife, Betty Meggers, and I did archeological work in Ecuador from 1954 to 1962, the Ecuadorian government generously permitted us to bring the duplicates from our excavations back to this country."

According to Evans, many museums persist in buying stolen antiquities even though they know they are perpetuating a scandalous situation. "If I were to show you pictures of looted graves in Peru," he asserts, "you wouldn't believe it. The graveyards look like they were bombed. There are huge craters where the looters have dug. The worst thieves of all are the gangs who come in at night in boats along the coast of Mexico carrying gas-powered Carborundum blade saws. They use these to cut off the carved facing of Mayan monuments so they can get away with a fragment of manageable size and weight. The unprotected Yucatán coast of Mexico makes it easy to carry out pillage such as this."

With the signing of a formal treaty with Mexico not long ago, the traffic in stolen Mayan art has dropped off slightly in the United States. If art dealers here are caught handling stolen material they can be forced to ship the objects back to Mexico without recompense. The market still exists elsewhere, however. Many of the biggest sales are taking place in Switzerland, West Germany, and Japan.

"Nevertheless," says Evans, "because of the international concern and the efforts of UNESCO to develop protective treaties, there appears to be some hope that the looting of archeological materials is slowing down. The world is increasingly beginning to appreciate the importance of archeology as a science that can assist in the reconstruction of past history. We can only hope that this enlightenment can somehow overcome the economic greed that motivates looting."

Mammals

Department of Vertebrate Zoology

The Smithsonian's zoological collections date back to 1850, when the Institution's first Secretary, Joseph Henry, hired the noted American naturalist Spencer Fullerton Baird. Congress had chartered the Smithsonian in 1846 as official custodian of natural history specimens belonging to the United States, and Baird made sure that the Institution became the clearinghouse for data and specimens sent back by U.S. exploring parties in the West.

Spencer Baird's leather-bound catalog of his personal collection, brought with him to the Smithsonian in 1850, is still on the shelves of the Mammals Division. In this ledger he noted nine basic items of information about each new animal that he acquired: catalog number, scientific name and sex of the animal, descriptive measurements, date and location of collection, who collected it, who prepared it, and when it was officially entered into the collection. For more than a century this continued to be the only information about its specimens that the Museum put on record. Then, in the late 1960s, computerization opened up a whole new era.

"With so many mammals in our collections, the problem of information retrieval was getting out of control," says Charles O. Handley, Jr., a curator of the division. "We had mammals in three acres of storage drawers, and thousands of new specimens arrived every year. It was grow-

The long-nosed nesogale, *like the* hemicentetes, *is an insectivore.*

The Magnificent Foragers

Three mammals of the Madagascaran jungle: a hedgehog, top; the spiny-furred hemicentetes, *center; and the* cheirogaleus amphijoroa, *bottom.*

ing more difficult to summarize information about the collection, and requests for information were increasing at the same time.

"For example," Dr. Handley points out, "somebody would write and ask, 'What do you have from Lawrence County, Ohio?' It would take hours of search to answer that question. Now we've developed a computer system that will almost instantly pick out our information on specimens from Lawrence County.

"The computer potential for data banking is open-ended. We still record the same nine items of information that Baird did, but now we can expand his one line to a whole block of data. We can put paragraphs or even pages of notes on field observations and laboratory analysis into the computer catalog, making it possible for us to do detailed research bearing on ecology and evolution that would have been impractical before."

One of Handley's projects is a study of 40,000 mammals collected in Venezuela. He has already fed more than 2 million lines of data on these collections into the computer—data about the ecology and geography of the region where the specimens were collected, sex and reproductive information, and details about the fleas, ticks, and other parasites found on the animals.

"Using our computer system, I can sort and correlate these data into innumerable combinations," he explains. "I can ask the computer, for example, what ticks are carried by a certain kind of bat in northern Venezuela, which life zones and elevations they inhabit, whether they occur also on other hosts. I can also ask for catalog numbers so that I can locate the best specimens in the collection."

Knowledge of the complex relationships between mammals and ticks, fleas, and other parasites is important to scientists because these parasites carry disease organisms that infect animals and humans. When a deadly viral fever broke out in Sierra Leone, the U.S. Public Health Service's Center for Disease Control turned to Henry W. Setzer, first of the Smithsonian's mammal researchers, to make large-scale use of the computer. He had sent collecting teams into the field in 16 African countries. For the most part ignoring the better documented larger animals, the teams concentrated instead on trapping shrews, hedgehogs, rabbits, foxes, weasels, skunks, gerbils, bats, rats, and mice—amassing over 50,000 small mammal specimens. Dr. Setzer consulted this collection and the computer data on it, and was able to quickly identify an African house rat as the source of the killer fever.

The Smithsonian's mammal collection now includes more than 500,000 specimens, wholly preserved in jars of alcohol or as dry skins and skeletal parts. Scientists constantly use the collection as a reference tool. For instance, a person may send a mammal to the Museum and request its exact identity. Researchers compare it with specimens in the collections. They note similarities or differences in the color, color pattern, quality of the fur, shape and size of the teeth, and the configuration of bones in the skull or skeleton.

"Medical researchers send us much of the material that we work on, but a lot comes in from the public, too," says Setzer. "A woman will send us an old bone that her little boy found out in the woods, and ask us to identify it. Over the years we've been shipped everything under the sun, from horse pelvis to whale skull. Sometimes these finds are important, adding to our knowledge of what kinds of animals may be found in a particular area."

American Indians and Eskimos

Department of Anthropology

The Smithsonian is the birthplace of anthropology as a profession in the United States. "The science was nurtured here in its early years before it became strongly established at American universities," states anthropologist William Sturtevant. The vigor of the Institution's efforts in this field was reflected in part by the growth of the collections. Material flowed in from all over the world, but the preponderance of it came from the American West and Arctic.

Explorations of Hopewell Indian burial mounds in the Midwest, Pueblo ruins and cliff dwellings in the Southwest, and other ancient sites yielded great quantities of archeological artifacts. The varied pattern of traditional American Indian and Eskimo life was documented by the accumulation of thousands of ethnological objects: bows and arrows, totem poles, canoes, clothing, baskets, blankets, pottery, textiles, and other Indian and Eskimo handicrafts began arriving at the Institution in profusion as early as the 1850s. Among the omnivorous collectors who sought out Indian study and display material were John Varden, who set up the Smithsonian's first Indian exhibits; Robert Kennicott, who made a one-man exploring trip by boat, canoe, and dog sled into the wilderness of the Canadian northwest; Major John Wesley Powell, explorer of the Colorado River, who collected a variety of artifacts from Indian country in the Great Basin area of the Southwest; and James Swan, who helped the Institution assemble a Northwest Coast collection that is among the finest in the world.

John C. Ewers, the Smithsonian's senior ethnologist and a specialist on Plains Indian history, is a relative of the legendary Buffalo Bill Cody. Dr. Ewers developed an interest in Indian culture as a young boy after becoming fascinated with Buffalo Bill's collection of Plains Indian artifacts. In the 1940s he and his family lived on a Blackfoot reservation in Montana where he was able to interview aging tribesmen who could still remember the buffalo hunts and tribal

A carved walrus tusk, made by Eskimos on Nunivak Island in the Alaskan Aleutians, depicts various animals of the Arctic and an unusual event—a rogue walrus killing a seal. This 20th-century, non-traditional piece was made for sale to tourists. The Navajo sandpainting replica, facing page, is one of 31 reproductions painted by Hastin Claw, or "Big Left-Handed," a Navajo Indian, in the 20th century. The black figure represents Father Sky, the blue figure Mother Earth.

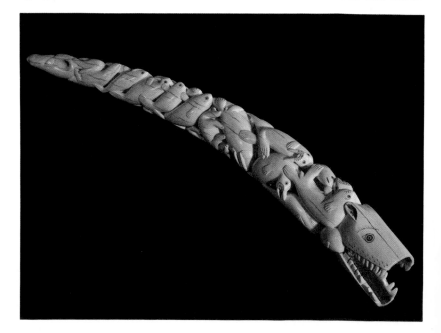

The Magnificent Foragers

wars of the 19th century. A classic study of the influence of the horse on Blackfoot culture grew out of Ewers' data.

The Institution still collects 19th-century material when it can, but it is much easier now to acquire more recent handicrafts. Though these are valuable both as a record of contemporary creativity and as a basis of comparison with the older material, the range of modern artifacts is diminishing as American Indians and Eskimos come to rely more and more on manufactured articles.

Meanwhile, the irreplaceable 19th-century specimens in the Smithsonian's collections still find new uses. "Some of our older collections are being carefully studied for the first time because their uniqueness is now recognized," Ewers points out. "For example, our

This atypically shaped late-19th-century polychrome water jar, below left, decorated with large birds, was collected at Acoma Pueblo in New Mexico. Above right, etched and darkened silhouettes on this ivory pipe with metal bowl depict a caribou hunt and other aspects of Eskimo daily life. From the Haida Indians of Queen Charlotte Island, British Columbia, comes this cedar box, below right. Carved and painted with a totemic design, it was collected by James Swan in the late 19th century.

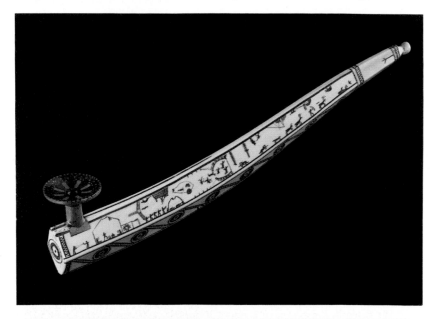

The Magnificent Foragers

large collection of Zuni pottery brought here in the 1880s is now being restored and restudied."

Earlier students of American Indian collections were concerned with primitive crafts and technology—methods of basket making, fire making, pottery making, and stoneworking. While interest in technology has persisted, there is now a greater interest in the range of variation in handicrafts and in the identification—where possible—of the works of individual artists and craftsmen, as well as in the social roles of these creative individuals in their societies.

One of the Smithsonian's outstanding early collections is from Arctic America: "Our collection of 19th-century Eskimo ethnographic material from Alaska is the most complete ever made, and is particu-

A Hupa Indian dance ornament, worn primarily in the hair. The 21-inch-long piece consists of three feather-tufted sticks bound to another large stick with a sharpened point.

Meanwhile, Back at the Museum

The first three drawings were done by Cheyenne Indian prisoners at Fort Marion in Ft. Lauderdale, Florida, in 1875. Making Medicine drew the top picture, Buffalo Meat the other two. Bottom, a later 19th-century Hopi Indian drawing of a kachina.

larly valuable because it was collected before there was any perceptible outside influence in the area," says Henry B. Collins, an authority on Eskimos and the Arctic regions. Most of this material was acquired in the 1870s and 1880s when the Smithsonian, upon invitation, nominated observers for the U.S. Signal Service in Alaska. Men such as E. W. Nelson, Lucien M. Turner, and John Murdoch were chosen to go north to make meteorological observations and to collect the little-known biological and ethnographic material of the far northern region. Their reports on housing, food, clothing, crafts, and other aspects of Eskimo life were subsequently published by the Smithsonian's Bureau of American Ethnology. Today, according to Dr. Collins, these reports are indispensable to students of life in the Arctic.

Collins' own field work in a series of buried villages along the Alaska coast and its offshore islands has brought extensive prehistoric Eskimo archeological collections to the Institution, providing one of the keys to the understanding of Eskimo cultural development.

Another important archeological collection—relating to the Plains Indians—has been assembled by Waldo Wedel. A native of Kansas, his work has taken him back regularly to his childhood neighborhood where he collected arrowheads and potsherds and wondered about the people who had left them there.

As a schoolboy Wedel had been intrigued by newspaper articles he read speculating whether the 16th-century Spanish explorer Coronado and his army of conquistadores had marched through Kansas on their futile search for the fabled riches of Quivira. Years later, at the site of a 16th-century Wichita Indian village, Wedel excavated rusty bits of Spanish chain armor, glass beads, and other European artifacts, compelling evidence that Coronado and other Spaniards had indeed come to central Kansas. An erstwhile question had been answered.

In addition to archeological remains, handicrafts, and other objects, the Smithsonian assembled an unparalleled collection of manuscripts on Indian and Eskimo religion, history, mythology, folklore, and social organization. These manuscripts are now preserved in the Institution's National Anthropological Archives along with a vast mass of Indian and Eskimo vocabularies, texts, and grammatical notes.

Many of these languages are now extinct, and the only way of learning anything about them is through studying this old material. In most cases, all that exists is a vocabulary of several hundred words and a scattering of grammatical information. "But it is still terribly important," notes Dr. Sturtevant. "A language is a language, worthwhile of study whether it be ancient Greek or American Indian. They're comparable products of human thought. Each tells you something new and different about humans."

Contributions to linguistic studies are being continued at the Smithsonian by scholars such as Ives Goddard, who is assembling data on Indian languages for a new 20-volume encyclopedia about native Americans that Sturtevant is editing; and by Robert Laughlin, who recently compiled the first dictionary of the Mayan language, Tzotzil, spoken by more than 125,000 Indians of the Mexican State of Chiapas. It is being used by scholars as a model for the creation of dictionaries of North American Indian languages. Moreover, its tremendous amount of detailed information on the meaning of words and

Joseph K. Dixon's photograph, top, of a mounted group of Crow, originally titled "Climbing the Western Slope," was taken on the Wannemaker expedition of 1912. The bottom photograph of Chief Medicine Crow was taken by Delancey Gill in 1913.

the way they are organized makes it useful to scientists who are trying to decipher ancient Mayan inscriptions.

Among the National Anthropological Archives' other documentary holdings are numerous drawings by Indian artists, including crayon and watercolor impressions they recorded of the Battle of Little Bighorn, and an incomparable collection of 50,000 photographs of Indians taken by early cameramen.

"Smithsonian scholars and the interested public find the photographs increasingly useful," adds the Archives' director, Herman Viola. "When Medicine Crow, chairman of the cultural committee of the Crow Indian Tribe, visited the Archives and examined the photo files, he found a wealth of data about his tribe that he did not know existed. He was able to identify and clarify the meaning of dozens of undocumented Crow Indian photographs."

Two discoveries in particular delighted Medicine Crow. He found a photograph of his grandfather, the famous Chief Medicine Crow, and a group portrait of his mother, father, and two brothers, taken in 1923.

"We're hopeful of many more such discoveries," says Dr. Viola. "And young Indians from many tribes now receive professional training so they can assemble and study tribal records. The Smithsonian provides copies of Archives photographs to Indian tribal archives and cultural centers. This is one of the ways in which a collection can serve as a living contemporary resource."

Small World

National Museum of Natural History

Little things mean a lot to Kjell Sandved, the Museum's photographer and cinematographer. Insects, for instance. He has filmed wasps and leafcutter ants in Panama, robber flies in Washington, D.C., and beetles in the Philippines. And innumerable moths, from Brazil to the South China Sea.

"One of my real loves," says Sandved, "is working with scientists to produce a documentary film oriented to a research project, focusing on a single biological event or concept." In doing so, he dramatically illuminates minute biological activities of the creatures he photographs. Scientists often cannot reconstruct such activities with the unaided eye, for they occur either too slowly, too rapidly, or are too complex. Photography, however, particularly with the help of macro lens close-ups and other devices, results in a permanent record that can be analyzed afterward at more convenient speed.

His film "Curious Creatures and Minimonsters," for example, documents facets of insect life from locomotion and mating to cryptic and rudimentary patterns of display, behavior, and evolution.

Smithsonian photographer Kjell Sandved prepares to film geometrids, noctuids, and hawk moths at Rancho Grande, Venezuela, a stop on their migratory route from South to North America.

The Magnificent Foragers

Index

Illustration and caption references appear in *italics*.

The Magnificent Foragers

The Magnificent Foragers

Photo Credits

Jacket: Paul J. Spangler
Front Matter: 1 (hermit crab) Raymond Manning; 2 (research vessel *Tunuyak*) William W. Fitzhugh; 3 (tree frog) Kjell B. Sandved.

I. The Tradition:
p. 12 Smithsonian Institution National Anthropological Archives; p. 13 (left) Karl V. Krombein; p. 13 (right) Smithsonian Archives; p. 14 Smithsonian Institution; p. 15 Smithsonian Archives; p. 16 John Bowden/Washington Star News Photo; p. 17-18 Smithsonian Archives; p. 19-20 Smithsonian Institution Anthropological Archives; p. 21-22 John Bowden/Washington Star News Photo; p. 23 (top) lent by John Kingsolver/USDA; p. 23 (bottom) Smithsonian Archives; p. 24 Smithsonian Archives; p. 25 John Bowden/Washington Star News Photo; p. 26 gift of Mrs. A. J. E. Davis, reproduced courtesy of the Hunt Institute for Botanical Documentation; p. 27 Smithsonian Institution Anthropological Archives.

II. The Solitary Researcher:
p. 29-35 Clyde F. E. Roper; p. 36 Mary Livingston Ripley; p. 37 (left, top right) Alice Tangerini; p. 37 (bottom right) S. Dillon Ripley; p. 38 Mary Livingston Ripley; p. 39 (top) Mary Livingston Ripley; p. 39 (bottom) S. Dillon Ripley; p. 41 S. S. Heaton; p. 42 (top) Donald R. Davis; p. 42 (bottom) Kjell B. Sandved; p. 43 Duckworth & Eichler; p. 44 Ernest A. Lachner; p. 45 (top) Ernest A. Lachner; p. 45 (middle) Ernest A. Lachner; p. 45 (bottom) Kjell B. Sandved; p. 46 (top) Kjell B. Sandved; p. 46 (bottom) Vitty & Seaborne, Ltd.; p. 47 Douglas Rogers; p. 48 Thomas McIntyre; p. 49 James G. Mead; p. 50 (left) James G. Mead; p. 50 (right) Roger Payne, courtesy of the National Geographic Society; p. 51 James G. Mead; p. 52 (left) George E. Watson; p. 52 (top right) George E. Watson; p. 52 (bottom right) Smithsonian Institution; p. 53 George E. Watson.

III. Two Islands:
p. 55 Smithsonian Institution; p. 56-58 George Silk; p. 59 (top) George Silk; p. 59 (bottom) Smithsonian Institution; p. 60 (top) George R. Zug; p. 60 (bottom) George Silk; p. 61 George Silk; p. 62 (left) George Silk; p. 62 (right) Richard W. Thorington; p. 63 Darelyn Handley; p. 64 Terry L. Erwin; p. 65 George Silk; p. 66 Neal Smith; p. 68-69 George Silk; p. 71 Kjell B. Sandved; p. 72 (top) James N. Norris; p. 73 James N. Norris; p. 74-75 Kjell B. Sandved; p. 76 James N. Norris; p. 77 Porter M. Kier; p. 79 Kjell B. Sandved.

IV. Looking at Culture:
p. 82 Ray Roberts Brown; p. 83 William H. Crocker; p. 84 (top) William H. Crocker; p. 84 (bottom) Ray Roberts Brown; p. 85-87 William H. Crocker; p. 88 Mary Livingston Ripley; p. 89 (left, top right) Victor Krantz/OPPS; p. 89 (bottom right) S. Dillon Ripley; p. 90 Mary Livingston Ripley; p. 91 Victor E. Krantz/OPPS; p. 92-95 Gordon D. Gibson; p. 97 Saul H. Riesenberg.

V. Plants of The Kingdom:
p. 99-104 Edward S. Ayensu; p. 105 Kjell B. Sandved; p. 106 Edward S. Ayensu; p. 107-108 Thomas R. Soderstrom; p. 109 (top) Robert W. Read; p. 109 (bottom) Edward S. Ayensu; p. 110-112 Robert W. Read; p. 113 Alice R. Tangerini; p. 114 (left) Alice R. Tangerini; p. 114 (right) J. Neff.

VI. Hard Stuff & Fiery:
p. 116 Tom E. Simkin; p. 117 Richard S. Fiske; p. 118 (left) Mike Harris; p. 118 (right) Dick Blank; p. 119 (top) Richard S. Fiske; p. 119 (bottom) William G. Melson; p. 120 (top) Richard S. Fiske; p. 120 (bottom) Donald W. Peterson; p. 121 Robert I. Tilling; p. 122 Tom E. Simkin; p. 123 Jeffrey B. Judd; p. 124-125 Richard S. Fiske; p. 126 Smithsonian Astrophysical Observatory; p. 127 Roy S. Clarke, Jr.; p. 128 Kurt A. Fredriksson; p. 129 (top) R. Cox; p. 129 (bottom) Robert F. Fudali;

p. 130 Lee Boltin; p. 131 (left, top right) Smithsonian Institution; p. 129 (bottom right) Dane Penland/OPPS; p. 132-133 Smithsonian Institution; p. 134 (top) Victor Krantz/OPPS; p. 134 (bottom) Brian H. Mason.

VII. Movable Seas & Oceans:
p. 137 courtesy of National Oceanic & Atmospheric Administration; p. 138 D. J. Stanley; p. 139 (top) Richard H. Benson; p. 139 (bottom) Ellie Benson; p. 140 Richard H. Benson; p. 141 Larry Isham; p. 142 Richard H. Benson; p. 143 Dan Budnik/Woodfin Camp; p. 144 Larry Isham; p. 145 Kjell B. Sandved; p. 146-147 Richard E. Grant.

VIII. Digs:
p. 149-152 Robert K. Vincent; p. 153 Susan Kaplan; p. 154 (top) Christopher Nagle; p. 154 (bottom) William W. Fitzhugh; p. 155 William W. Fitzhugh; p. 156 Robert K. Vincent; p. 157 (left top, left center, left bottom) Victor Krantz; p. 157 (right top) Robert K. Vincent; p. 158 Robert K. Vincent; p. 159 Kjell B. Sandved; p. 160 (left) Donald Ortner; p. 160 (right) Kjell B. Sandved; p. 161 (top) Kjell B. Sandved; p. 161 (bottom) Donald Ortner; p. 162 (left) Douglas H. Ubelaker; p. 162 (right) Rick Madden; p. 163 (left) Kjell B. Sandved; p. 163 (right) Douglas H. Ubelaker; p. 164-165 J. Lawrence Angel.

IX. The Ice Ages & Before
p. 167 Allison Witter, courtesy of the National Geographic Society; p. 168-171 Victor Krantz/OPPS; p. 172 (top) Larry Isham; p. 172 (bottom) Erle Kauffman; p. 173 (left top, left bottom) Donald Dean; p. 173 (bottom right) Kjell B. Sandved; p. 174 Kenneth M. Towe; p. 175-176 Robert J. Emry; p. 177 Smithsonian Institution; p. 178 Francis Hueber; p. 179 (top) James A. Doyle; p. 179 (center left, center right, bottom) Leo J. Hickey.

X. Amateuris populis:
p. 181 Pamela Meyer; p. 182 (left) Kjell B. Sandved; p. 182 (right) Pamela Meyer; p. 183 Kjell B. Sandved; p. 184 (left) Gary F. Hevel; p. 184 (right) Kjell B. Sandved; p. 185 Chip Clark; p. 186 (left top, left center, left bottom) Karl V. Krombein; p. 186 (right) P. B. Karunaratne.

XI. The Practical Naturalist:
p. 189 Paul D. Hurd, Jr.; p. 190 Mason E. Hale; p. 191 Kjell B. Sandved; p. 192 Jack W. Pierce.

XII. Meanwhile, Back at the Museum:
p. 194 Joseph Rosewater; p. 195 Raymond Bouchard; p. 196-197 Kjell B. Sandved; p. 198 Duane Hope; p. 199 (top) Kjell B. Sandved; p. 199 (center, bottom) Roger F. Cressey; p. 200 Kjell B. Sandved; p. 201 Victor G. Springer; p. 202 Alice Tangerini; p. 203 Joan Nowicke; p. 204 Kjell B. Sandved; p. 205 George R. Zug; p. 207 Victor Krantz; p. 208-209 H. Uible; p. 210-213 Victor Krantz; p. 214 Kjell B. Sandved; p. 215 Joseph Dixon/Smithsonian Institution National Anthropological Archives; p. 216 Kjell B. Sandved.